Sonja Knapp

Plant Biodiversity in Urbanized Areas

VIEWEG+TEUBNER RESEARCH

Sonja Knapp

Plant Biodiversity in Urbanized Areas

Plant Functional Traits in Space and Time,
Plant Rarity and Phylogenetic Diversity

With forewords by Dr. Stefan Klotz and Prof. Dr. Rüdiger Wittig

VIEWEG+TEUBNER RESEARCH

Bibliographic information published by the Deutsche Nationalbibliothek
The Deutsche Nationalbibliothek lists this publication in the Deutsche Nationalbibliografie;
detailed bibliographic data are available in the Internet at http://dnb.d-nb.de.

Dissertation Universität Frankfurt, 2009

D 30

1st Edition 2010

Reader: Dorothee Koch | Anita Wilke

Vieweg+Teubner is part of the specialist publishing group Springer Science+Business Media.
www.viewegteubner.de

Cover design: KünkelLopka Medienentwicklung, Heidelberg
Printing company: STRAUSS GMBH, Mörlenbach
Printed on acid-free paper
Printed in Germany

ISBN 978-3-8348-0923-0

Foreword

Urban areas are increasing more and more and human's first contact to nature will take place in cities. More than 50 % of the world's human population is concentrated in urban areas; this number is even higher in Central Europe. These are the main reasons for the increasing number of studies on urban ecology including urban flora and vegetation. One surprising outcome of these studies was the higher species richness in urban areas in comparison to the open cultural landscape in Central Europe. This stable pattern has been found within several studies on several cities since the 70s of the last century.

The main tasks of the dissertation of Sonja Knapp can be summarised in the following questions: What are the main causes for higher species richness and what processes govern this pattern? Is species richness linked with ecological traits and is species richness in urban areas linked with phylogenetic diversity?

Sonja Knapp's dissertation "Plant Biodiversity in Urbanized Areas – Analyses of Plant Functional Traits in Space and Time, Plant Rarity and Phylogenetic Diversity" presents new insights into biodiversity processes in urban areas. First, the trait composition within urban floras is significantly different from non-urban floras; second, urbanization threatens rare native species, while common native species and aliens profit from urban land use; third, a clear trait shift was found within a time series of floras (over more than 300 years) of the city of Halle (Saale) comparable to spatial differences between urban and non-urban landscapes; fourth, it was shown that the phylogenetic diversity of urban areas does not reflect their high species richness. The reduced phylogenetic information might decrease the flora's capacity to respond to environmental changes.

Sonja Knapp is not only dealing with trait composition and phylogenetic diversity of floras but also with the influence of different spatial and temporal scales. Spatial differences between urban and non-urban areas are reflected in time series of urban floras in the same way. Spatial patterns of urban floras depend on scales. Results of such studies must be set into the scales context.

Sonja Knapp's work is an outstanding contribution to biodiversity studies in urban areas. Therefore, I am sure that this publication will be of interest and great benefit not only for urban ecologists but also for ecologist in general and environmental managers, city and landscape planners. Additionally, the basic findings have a high value for the development of biodiversity indicators for urban areas. To sum up, this publication is a highly substantial contribution to urban ecology.

Stefan Klotz

Foreword

Today, cities are not only the most important habitat of humans but they also host an astonishingly great number of plant species. Regarding Central Europe, the number of plant species in cities is very often significantly higher than in rural areas with comparable sizes. A steadily increasing urbanization leads to the following interesting questions: What traits enable species to survive in the cities? What does this multitude of species contribute to the maintenance of the world-wide biodiversity? These questions can best be answered by comparing the traits of urban species with the traits of species from the cities' surrounding or from rural areas.

The dissertation of Sonja Knapp "Plant Biodiversity in Urbanized Areas – Analyses of Plant Functional Traits in Space and Time, Plant Rarity and Phylogenetic Diversity" is based exactly on these comparisons – carried out on different scale ranges.

Analyses like the one made by Sonja Knapp could be conducted only in the younger past as the crucial requirement – the availability of datasets of traits of whole florae – is just in its early stages. In the field of current urban ecological research, the dissertation of Sonja Knapp is, therefore, pioneer work in the true sense of the meaning.

Sonja Knapp is not only dealing with plant traits but also with plants' rareness and their phylogenetic diversity. Accordingly, her work has to be taken as an important contribution to the question of the importance of cities for biodiversity. Therefore, I am convinced that this publication will be of great interest and benefit to all ecologists who are interested in the reaction of organisms to changing environments, in biodiversity problems, and, of course, in urban flora.

Rüdiger Wittig

Preface

Several people at the Helmholtz Centre for Environmental Research – UFZ in Halle (Saale) and the Johann Wolfgang Goethe University in Frankfurt (Main) accompanied and supported my work on this thesis. In the first place, I want to thank my supervisors Prof. Dr. Rüdiger Wittig and Dr. Stefan Klotz as well as Dr. Ingolf Kühn who always lent me a helping hand. Thanks to Jun.-Prof. Dr. Oliver Tackenberg for reviewing the thesis, to Dr. Christine Römermann who contributed her knowledge, help, and good mood, and to Dr. Oliver Schweiger for his help and expertise. Thanks to all those whose work provided the basis for my studies (loads of data); many thanks to all my colleagues who made the time of my thesis a good one and to my family who never let me down!

At the time this book was published, several of its chapters had already been published separately in scientific journals, but in different forms: Each article has to be understandable on its own, but the chapters of this book can and should refer to each other. Therefore, the texts of the chapters partly deviate from the texts of the associated articles. I want to thank Wiley-Blackwell for the permission to reprint parts of the articles from "Diversity and Distributions" and "Ecology Letters" and the Czech Botanical Society to reprint a figure from the article in "Preslia". The articles that correspond to the chapters of this book shall be named here:

Chapter I: Knapp S., Kühn I., Wittig R., Ozinga W.A., Poschlod, P. & Klotz S. (2008). Urbanization causes shifts in species' trait state frequencies. Preslia 80, 375-388

Chapter III: Knapp S., Kühn I., Bakker J.P, Kleyer M., Klotz S., Ozinga, W.A., Poschlod, P., Thompson, K., Thuiller, W. & Römermann C. How species traits and affinity to urban land use control large-scale species frequency. Diversity and Distributions 15, 533-546

Chapter V: Knapp S., Kühn I., Schweiger O. & Klotz S. (2008). Challenging urban species diversity: contrasting phylogenetic patterns across plant functional groups in Germany. Ecology Letters 11, 1054-1064

Sonja Knapp

Summary

Although urban areas only occupy *c.* 2.8% of the earth's land surface, urbanization threatens biodiversity since areas of high human population density often coincide with high biodiversity: Cities harbor more species than their rural surroundings, at least over large enough scales. However, species richness does not necessarily cover all aspects of biodiversity such as the functional composition of species assemblages, species rarity or phylogenetic relationships. Ignoring these aspects of biodiversity, our understanding of how species assemblages develop and change in a changing environment remains incomplete. Moreover, knowledge of these aspects can give valuable information for the conservation of species diversity. In this study, vascular plant biodiversity of urbanized areas in Germany was analyzed with respect to these three aspects using statistical methods: Extensive databases on occurrence, traits, and relatedness of species (FLORKART, BiolFlor, LEDA) and on land use in Germany (Corine Landcover) provided the data to answer the following questions:

Does the functional composition of floras differ between urbanized and rural areas in Germany? (Chapters I and II)

Does the affinity of a plant species to urban land use together with its traits influence the rarity of the species? (Chapter III)

Does the functional composition of an area's flora change over time when the area becomes more and more urbanized? (Chapter IV)

Are the floras of urbanized areas phylogenetically more diverse than the floras of rural areas and does the phylogenetic diversity of plant species characterized by a specific functional trait differ between urbanized and rural areas? (Chapter V)

(1) Urbanization is one of the most extreme forms of land transformation. It deeply changes the structure and characteristics of a landscape. The specific characteristics of urban environments such as increased temperatures, reduced air moisture or a high degree of fragmentation, do not allow all species of the respective regional species pool to exist in urbanized areas. The according species turnover along urbanization gradients is supposed to change the frequencies of species trait states in species assemblages: In Chapter I, we hypothesized that the vascular flora of urbanized and rural areas in Germany differs in the frequency of trait states, and asked what traits enable a plant to cope with urban conditions. We performed the analyses on the basis of grid-cells sized 130 km^2 each. These grid-cells were divided into urbanized, agricultural, and forested/semi-natural ones, depending on the percentage of land use in the cell. Chapter I showed that urbanized grid-cells have e.g. higher proportions of wind-pollinated plant species, species with scleromorphic or overwintering green leaves or species dispersed by animals than non-urbanized grid-cells. Vice versa, urbanized grid-cells had e.g. less species that were pluriennial, dispersed by wind or had hygromorphic leaves than non-urban grid-cells. Climate, disturbance and fragmentation were, amongst others, discussed as possible drivers of these patterns. The results show that shifts in land use can change the trait state composition of plant assemblages. Furthermore, far-reaching urbanization is able to homogenize our flora with respect to trait state frequency.

However, patterns of biodiversity can change with scale: Cities are richer in species than rural areas on national scales but a city center can be poorer in species than an area of the same size in the near countryside. Large-scale analyses give important information on biodiversity but can also yield wrong conclusions for e.g. local nature conservation. In Chapter II, we tested the hypothesis of Chapter I for a smaller scale: We compared plant trait state frequency in protected areas and randomly selected 0.06 km^2-plots in the city of Halle and its rural environs in Central Germany. The functional composition of urban and rural plant assemblages was less different on the small than on the large scale. There were similarities between scales such as increased proportions of animal-dispersed species and therophytes in the urban study sites. Other patterns differed such as increased proportions of pluriennials or species with hygromorphic leaves in Halle's urban study sites. Local conditions can mitigate effects of urban land use, such as forests or parks within a city that support plants that are less frequent in typical urban built-up environments. Consequently, semi-natural habitats in an urban area contribute importantly to the conservation of biodiversity in cities. However, also the stochastic lack of rare plant species in small study sites can cause differences between large- and small-scale patterns.

(2) Urbanization especially threatens rare native plant species, while alien and common native species may profit from urban land use. An understanding of why species are rare is necessary to develop effective conservation strategies. Both plant traits and extrinsic threats such as urbanization can contribute to species rarity. In Chapter III, we assessed how the affinity of plants to urban land use interacts with plant traits in determining species frequency in Germany. Common species had a high affinity to urban land use in contrast to rare species. The affinity to urban land use interacted with the species' type of pollination, type of reproduction, habitat preferences regarding temperature and moisture and with the existence of hemirosettes. This means that many rare species, especially those depending on biotic pollinators, those reproducing vegetatively or those preferring cool, moist or non-ruderal habitats might already have disappeared from urbanized areas. Consequently, cool, moist and rarely disturbed habitats of older successional stages, e.g. alluvial forests, can especially contribute to the conservation of rare species in urbanized areas. Chapter III showed the potential of analyses combining traits and environmental effects for understanding the causes of rarity to derive better conservation strategies.

(3) Today's state of biodiversity is the product of recent environmental conditions but also reflects historical developments. Documents on historical floras provide unique opportunities to analyze past changes, to show trends in biodiversity, and to explain the structure of recent floras. In Chapter IV, we studied the development of the flora of Halle in the last 320 years. More than 20 floras, the earliest dating back to the year 1687, provided information on plant occur-

rences. According to the spatial analyses in chapters I and II, urbanization should also change the species composition and functional composition of floras when increasing over time. Indeed, there was a species turnover of 22% within the study period, accompanied by a turnover in traits: The proportion of natives and archaeophytes decreased while the proportion of neophytes increased. Species of bogs, nitrogen-poor habitats or plants with hydromorphic leaves got extinct more often than expected by chance. Species dispersed by animals, plants preferring nitrogen-rich or hot habitats, and phanerophytes were, amongst others, overrepresented among introduced species. Transformations from agriculture to urban land use, drainage of bogs, climatic changes, contamination of habitats, and gardeners' preferences for specific plants were discussed as main drivers of these developments. Chapter IV showed that changes in the functional composition of Halle's flora already started centuries ago.

(4) Given the high vascular plant species richness of urbanized areas in Germany, Chapter V investigated whether these also have a higher phylogenetic diversity than rural areas, and whether phylogenetic diversity patterns differ systematically between species groups characterized by specific functional traits. Chapter V showed that phylogenetic diversity of urban areas does not reflect their high species richness. However, species that can cope with urban land use such as plants with scleromorphic leaves or self-pollinated species have a higher urban than rural phylogenetic diversity. Contrarily, plants with trait states less suitable for urban land use, e.g. plants with hygromorphic leaves or pluriennials, have a higher rural than urban phylogenetic diversity. Hence, high urban species richness is mainly due to more closely related species that are functionally similar and able to deal with urbanization. This diminished phylogenetic information might decrease the flora's capacity to respond to environmental changes.

The results were discussed in the context of other studies on urban ecology in the Chapter "Synthesis and Conclusions". The present study revealed effects of urban land use on the structure of vascular plant species assemblages, gave insights into the development of urban species assemblages, and provided important background information for the conservation of plant biodiversity in urbanized landscapes. It showed that urbanization is able to globally decrease plant diversity by fostering common species with "urban-adapted" trait states, by limiting rare species with traits less suitable for urban environments, and by decreasing phylogenetic diversity. The study also provided evidence that semi-natural areas within cities, e.g. protected areas, parks, and even gardens, can protect species with traits less suitable for the typical urban built-up environment. Therefore, a mix of semi-natural habitats and typical urban-industrial habitats within a city (if the latter are allowed for spontaneous vegetation) should permit both semi-natural and typical urban floristic elements and thus enable functional and phylogenetic diversification. Because more than half of the total world population lives in cities, urban nature is of special importance for global biodiversity: People who appreciate nature are probably willing to protect it. As for many urban dwellers, urban nature is the only nature they can experience everyday, it is also the only kind of nature that can show them the values of biodiversity.

Zusammenfassung

Städte bedecken nur 2,8% der Landflächen der Erde. Urbanisierung findet allerdings häufig in Gegenden statt, die sich durch besonders hohe Artenzahlen auszeichnen und stellt daher eine weltweite Bedrohung für die Biodiversität dar. Artenzahlen sind aber nur ein Teilaspekt der Biodiversität und lassen detailliertere Maße, wie zum Beispiel die funktionelle Zusammensetzung von Artengemeinschaften, die Seltenheit von Arten oder ihre verwandtschaftlichen Beziehungen, außer Acht. Diese Maße können Aufschluss darüber geben, wie Artengemeinschaften entstehen und wie sie auf Veränderungen ihrer Umwelt, beispielsweise auf Urbanisierung, reagieren. Sie können damit auch Hinweise für einen effektiven Artenschutz geben. In der vorliegenden Dissertation wurden diese drei Aspekte der Diversität der Gefäßpflanzen in Städten am Beispiel Deutschlands und anhand statistischer Methoden untersucht. Dabei konnten mit Hilfe umfangreicher Datenbanken zu Vorkommen, Merkmalen und Verwandtschaftsgrad der Arten (FLORKART, BiolFlor, LEDA) sowie zur Landnutzung in Deutschland (Corine Landcover) folgende Fragen beantwortet werden:

Unterscheidet sich die funktionelle Zusammensetzung der Flora städtischer Gebiete von derjenigen der Flora in ländlichen Gebieten? (Kapitel I und II)

Beeinflusst die Affinität einer Pflanzenart zu städtischer Landnutzung in Interaktion mit den Merkmalen der Art ihre Seltenheit? (Kapitel III)

Führt die Urbanisierung eines Gebietes über einen längeren Zeitraum hinweg zu Veränderungen in der funktionellen Zusammensetzung der Flora des Gebietes? (Kapitel IV)

Sind die Floren städtischer Gebiete phylogenetisch (verwandtschaftlich) diverser als die Floren ländlicher Gebiete und unterscheidet sich die phylogenetische Diversität von Artengruppen, die durch bestimmte Merkmale gekennzeichnet werden, zwischen städtischen und ländlichen Gebieten? (Kapitel V)

(1) Urbanisierung verändert die Struktur und Eigenschaften von Landschaften grundlegend. Die spezifischen Eigenschaften städtischer Landschaften, beispielsweise die im Vergleich zum Umland erhöhten Temperaturen, die verringerte Luftfeuchte oder der hohe Fragmentierungsgrad, führen dazu, dass nicht alle Arten des jeweiligen regionalen Artenpools in Städten vorkommen. Der Wandel des Artenspektrums (Arten-*Turnover*) entlang eines Stadt-Land-Gradienten sollte auch Veränderungen in der funktionellen Zusammensetzung der Artengemeinschaften (Merkmals-*Turnover*) nach sich ziehen. Ausgehend von dieser Hypothese wurde in Kapitel I untersucht, ob bestimmte Pflanzenmerkmale in den Floren städtischer Gebiete Deutschlands häufiger oder seltener vorkommen als in den Floren ländlicher Gebiete. Die Analyse fand auf Basis von Rasterzellen à 130 km² statt, die in städtisch dominierte, landwirtschaftlich dominierte und forstlich/naturnah dominierte Rasterzellen eingeteilt wurden, entsprechend des prozentualen Anteils der jeweiligen Landnutzungstypen. Es zeigte sich, dass in städtischen Gebieten u.a. mehr Pflanzen vorkommen, die windbestäubt sind, skleromorphe oder überwinternd grüne Blätter haben oder von Tieren verbreitet werden als in ländlichen Gebieten. Umgekehrt sind beispielsweise mehrjährige Arten, windverbreitete Arten oder Arten mit hygromorphen Blättern in städtischen Gebieten seltener als in ländlichen Gebieten. Als mögliche Ursachen für diesen Merkmals-*Turnover* wurden u. a. Klima, Störungsgrad und Fragmentierung der Städte diskutiert. Den Ergebnissen zufolge kann eine weitreichende Urbanisierung der Landschaft tatsächlich die Merkmalsanteile betroffener Pflanzengemeinschaften verschieben und eine funktionelle Homogenisierung der Flora nach sich ziehen.

Allerdings können solche Biodiversitätsmuster mit der Betrachtungsebene variieren: Auf einer nationalen Skala sind Städte meist artenreicher als ländliche Gebiete; ein Stadtzentrum kann aber deutlich artenärmer sein als ein Gebiet gleicher Größe im Umland der Stadt. Daher wurde in Kapitel II die gleiche Hypothese getestet wie in Kapitel I, aber auf einer kleineren Skala: Schutzgebiete und zufällig ausgewählte, 0.06 km² große Untersuchungsflächen in der Stadt Halle in Sachsen-Anhalt wurden mit entsprechenden Flächen der benachbarten

Landkreise Saalkreis und Mansfelder Land hinsichtlich der funktionellen Zusammensetzung ihrer Floren verglichen. Insgesamt traten auf dieser kleineren Skala deutlich weniger Unterschiede in der funktionellen Zusammensetzung der Stadt- und Landflora auf als auf der Skala aus Kapitel I. Den Ergebnissen des ersten Kapitels entsprachen zum Beispiel der erhöhte Anteil von tierverbreiteten Arten und von Therophyten in den städtischen Untersuchungsgebieten; im Gegensatz zur größeren Skala gab es in Halle aber u. a. mehr Pflanzen mit hygromorphen Blättern und mehrjährige Arten als im ländlichen Umland. Demzufolge können lokale Bedingungen die Effekte städtischer Landnutzung mildern: Innerstädtische Wälder, aber auch Parks und Gärten können Arten Lebensraum bieten, die ansonsten in Städten eher selten sind. Naturnahe Gebiete in einer Stadt können demzufolge wesentlich zum Schutz der Biodiversität beitragen. Die Unterschiede zwischen den Skalen können aber auch dadurch zustande kommen, dass seltene Arten auf großen Flächen mit höherer Wahrscheinlichkeit vorkommen als auf kleinen Flächen, also in den Rasterzellen vorhanden sind, in den relativ kleinen Untersuchungsgebieten von Kapitel II aber nicht.

(2) Es sind insbesondere seltene einheimische Pflanzenarten, die durch Urbanisierung bedroht werden; Neophyten und häufige einheimische Arten können sogar davon profitieren. Um effektive Maßnahmen zum Schutz der Biodiversität entwickeln zu können, ist ein Verständnis der Mechanismen, die Seltenheit bedingen, unablässig. Neben externen Einflussfaktoren, wie der Urbanisierung, kann Seltenheit auch von den Merkmalen einer Art abhängen. In Kapitel III wurde daher untersucht, wie sich die Merkmale einer Pflanzenart zusammen mit ihrer Affinität zu städtischer Landnutzung auf ihre Seltenheit auswirken. Dabei zeigten häufige Arten generell eine hohe Affinität zu städtischer Landnutzung, seltene Arten zeigten eine geringe Affinität. Die Stadtaffinität interagierte mit den Ansprüchen der Arten an Temperatur und Feuchtigkeit sowie mit der Art ihrer Fortpflanzung, ihrer Bestäubungsart und der Ausbildung einer Hemirosette. Das bedeutet, dass insbesondere die Arten aus Städten verschwinden, die in kühlen oder feuchten Habitaten leben, sich vegetativ fortpflanzen, biotisch bestäubt werden oder nicht mit ruderalen Bedingungen zurechtkommen. Folglich

kann das Vorhandensein entsprechender Habitate in Städten, beispielsweise von Auenwäldern, die relativ kühl, feucht und ungestört sind und zudem ein fortgeschritteneres Sukkzessionsstadium repräsentieren als viele typisch urbanindustrielle Habitate, zum Schutz seltener Arten in städtisch geprägten Landschaften beitragen. In Kapitel III konnte außerdem gezeigt werden, dass Analysen, die Pflanzenmerkmale und Umwelteinflüsse kombinieren, ein großes Potential für das Verständnis von Seltenheit haben und somit die Möglichkeit bieten, Maßnahmen zum Schutz seltener Arten abzuleiten.

(3) Der heutige Zustand von Artengemeinschaften hängt neben den heutigen Umweltbedingungen auch von zeitlich zurückliegenden Entwicklungen ab. Anhand historischer Florenwerke kann man zurückliegende Veränderungen einer Flora nachvollziehen, Entwicklungstrends offenlegen und die Struktur heutiger Floren erklären. In Kapitel IV wurde die funktionelle Zusammensetzung der Flora der Stadt Halle für den Zeitraum zwischen 1687 und 2008 untersucht. Mehr als 20 Florenwerke informierten über das Vorkommen der Pflanzenarten. Entsprechend des in den Kapiteln I und II für räumliche Stadt-Land-Gradienten gezeigten Merkmals-*Turnover*, sollte auch die über den Untersuchungszeitraum hinweg zunehmende städtische Landnutzung im Untersuchungsgebiet Änderungen in der funktionellen Zusammensetzung der Flora nach sich ziehen. Tatsächlich fanden im Untersuchungszeitraum sowohl ein Arten- als auch ein Merkmals-*Turnover* statt. Der Arten-*Turnover* betrug zwischen 1687 und 2008 22%. Der Anteil der einheimischen Arten und der Archäophyten in der Flora nahm in diesem Zeitraum deutlich ab; der Anteil der Neophyten nahm entsprechend zu. Der Vergleich der Merkmale der 1687 in Halle existierenden Arten mit denen der seitdem ausgestorbenen Arten zeigte, dass Arten der Moore, der stickstoffarmen Habitate und Pflanzen mit hydromorphen Blättern häufiger ausgestorben sind als erwartet. Der entsprechende Vergleich der Flora von 1687 mit den seitdem eingewanderten Neophyten zeigte, dass u. a. Arten, die von Tieren verbreitet werden, Arten stickstoffreicher und sehr warmer Habitate sowie Phanerophyten häufiger eingewandert sind als erwartet. Als Hauptverursacher dieser Entwicklungen wurden die Umwandlung landwirtschaftlicher in städtische Habitate, die

Entwässerung von Mooren, klimatische Veränderungen, Habitatverschmutzung sowie die bevorzugte Anpflanzung bestimmter Arten in Gärten und anderen städtischen Grünanlagen diskutiert. Die Ergebnisse zeigen, dass die Zunahme der städtischen Landnutzung im Untersuchungsgebiet bereits seit mehreren Jahrhunderten Veränderungen in der funktionellen Zusammensetzung der Flora nach sich zieht.

(4) Angesichts der Tatsache, dass städtisch geprägte Gebiete in Deutschland artenreicher sind als ländliche Gebiete, wurde in Kapitel V untersucht, ob erstere auch phylogenetisch diverser sind und ob sich die phylogenetische Diversität von Pflanzenarten, die ein bestimmtes Merkmal teilen, systematisch zwischen Stadt und Land unterscheidet. Dazu wurden die in Kapitel I definierten städtisch, landwirtschaftlich und forstlich/naturnah dominierten Rasterzellen hinsichtlich der phylogenetischen Diversität ihrer Flora verglichen. Es zeigte sich, dass die Flora in städtischen Gebieten, obwohl artenreicher, nicht phylogenetisch diverser ist als die Flora ländlicher Gebiete. Stattdessen ist die phylogenetische Diversität in städtischen Gebieten sogar etwas niedriger als in ländlichen Gebieten. Allerdings sind solche Artengruppen in Stadtregionen phylogenetisch diverser, die durch „stadtgeeignete" Merkmale charakterisiert sind, zum Beispiel Pflanzen mit skleromorphen Blättern oder selbstbestäubte Arten. Umgekehrt sind Artengruppen, die weniger „stadtgeeignete" Merkmale teilen, zum Beispiel Pflanzen mit hygromorphen Blättern oder mehrjährige Arten, in Stadtregionen phylogenetisch weniger divers als auf dem Land. Folglich basieren die hohen Artenzahlen deutscher Städte vor allem auf relativ eng verwandten Arten, die „stadtgeeignete" Merkmale teilen und gut mit den städtischen Umweltbedingungen zurechtkommen. Eine Reduzierung der phylogenetischen Diversität unserer Flora, beispielsweise aufgrund von Urbanisierung, könnte ihre Reaktionsfähigkeit gegenüber Veränderungen der Umwelt reduzieren.

Die Ergebnisse wurden im Kapitel „Synthesis and Conclusions" zusammenfassend diskutiert und in den Kontext anderer stadtökologischer Studien gestellt. Insgesamt konnte in dieser Arbeit gezeigt werden, wie städtische Landnutzung die Struktur von Pflanzengemeinschaften beeinflusst und welche Mechanismen

die Zusammensetzung von Artengemeinschaften in Städten steuern. Den Ergebnissen zufolge trägt Urbanisierung zu einem globalen Rückgang der Phytodiver-
sität bei, indem sie häufige Arten, die gut an städtische Landnutzung angepasst
sind, fördert, seltene Arten, die weniger gut an städtische Landnutzung angepasst
sind, zurückdrängt und die phylogenetische Diversität von Pflanzengemeinschaften reduziert. Die Ergebnisse legen außerdem nahe, dass insbesondere
naturnahe Habitate in Städten, wie zum Beispiel Naturschutzgebiete, aber auch
Parks und Gärten, Arten Lebensraum bieten können, die in der bebauten Stadt-
matrix nicht oder nur selten vorkommen. Innerhalb einer Stadt kann folglich eine
Kombination solcher naturnaher Habitate mit typisch städtischen Habitaten (vor-
ausgesetzt in letzteren wird eine spontane Besiedelung durch Pflanzen zugelas-
sen) zum Erhalt einer funktionell und phylogenetisch vielfältigen Stadtnatur
beitragen. Da mehr als die Hälfte der Weltbevölkerung in Städten lebt, ist eine
vielfältige Stadtnatur von besonderer Bedeutung für die Artenvielfalt weltweit:
Wer etwas wertschätzen lernt, ist wahrscheinlich auch bereit, es zu schützen. Da
die Stadtnatur aber für viele Bewohner die einzige Natur ist, die sie tagtäglich
erleben können, ist sie auch die einzige, an der sie die den Wert der Vielfalt der
Natur erleben und begreifen können.

Contents

List of Figures

List of Tables

General Introduction

1. World Urban Population Development

The world is urbanizing rapidly. In the year 1900, the total world population was around 1.6 billion people, of which 13% lived in cities (United Nations 2003; United Nations 2006). Within 105 years, the total world population increased to 6.46 billion people and the percentage of urban dwellers increased to 50% (United Nations 2006). This development is projected to continue to a total world population of 8.2 billion people and 60% of urban dwellers in 2030 (United Nations 2006; Fig. I1). In Europe and Germany, the percentage of people living in urban areas is even higher, with 72% and 75% respectively (United Nations 2006). Naturally, the increase in urban population was and still is accompanied by an increase in urban land use. In Germany, settlement and transport areas cover around 13% of the total land surface (Umweltbundesamt *et al.* 2007). Although this is far behind agricultural (53%) and forest land cover (30%), urban land use is the land-use type with the highest growth rate: Around 114 ha of land in Germany are transformed to settlement or transport area every day (Umweltbundesamt *et al.* 2007).

2. Urban Ecology

Regardless of the importance of cities for human civilization (or because of their importance as anthropogenic centers), biologists neglected them for a long time. Cities were claimed to be purely artificial and not interesting for biological research (McDonnell 1997). Although one of the earliest urban biological studies dates back to the year 1866 (Nylander 1866, cited in Wittig 2002), it was not until the mid of the 20th century that urban nature attracted the attention of a broader biological audience. As biologists began to notice urban nature, the first thing to do was to inventory the flora and fauna of cities (for a review see Wittig

2002). It then appeared that urban species assemblages are not random but that specific plant species grow in cities while others do not (Wittig & Sukopp 1998).

In the present study, "urban ecological analyses" are defined as analyses that deal with the ecosystems of urbanized areas, and not analyses that deal with urban planning or urban development. Thus, the study follows the definition of urban ecology in a strict sense by Wittig and Sukopp (1998).

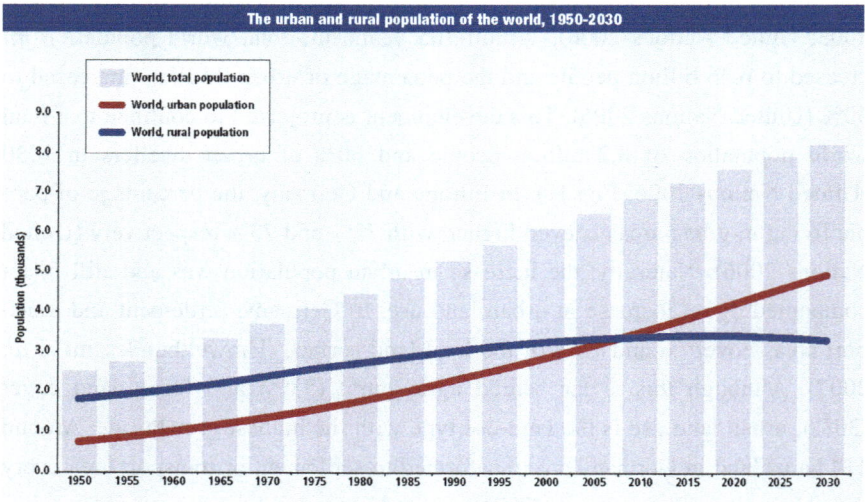

Figure I. 1 – World population development 1950-2030
Urban and rural population numbers are indicated separately. Population numbers from 2005 onwards are projected. Figure by the United Nations Department of Economic and Social Affairs, Population Division. http://www.un.org/ esa/population/unpop.htm, accessed on 04th of April 2007.

3. Urban Biodiversity

3.1. Species Richness

A surprising fact researchers first noticed when comparing urban and rural floras in the 1970ies is the high species richness of cities that often exceeds the richness of the surrounding countryside (Walters 1970; Haeupler 1975). This pattern was confirmed for vascular plants and several animal taxa in Europe (Pyšek 1993; Araújo 2003; Kühn *et al.* 2004a; Wania *et al.* 2006), North America (Dobson *et al.* 2001; McKinney 2002; Hope *et al.* 2003), South America (Fjeldså & Rahbek 1998), and sub-Saharan Africa (Balmford *et al.* 2001). The higher urban species richness in both native and foreign species can be explained because, firstly, many cities developed in geologically and structurally heterogeneous landscapes (Fig. A1 in the appendix). Obviously, the high geological diversity is not due to urban land use; rather, cities were often founded in geologically rich, and thus also species rich areas because these are favorable for human requirements (Kühn *et al.* 2004a). Secondly, cities are highly structured themselves with many different habitat- and land-use types in a small area, giving room to a variety of species (Sukopp 1998; Niemelä 1999). The high urban heterogeneity especially contrasts with the homogeneity of many modern agricultural landscapes. These are often dominated by the same crop and treatment over a wide area. Thirdly, cities often include various habitat types that are rare or absent in the surroundings (Sukopp & Starfinger 1999; Godefroid & Koedam 2007). Fourthly, the high urban temperatures, which in temperate climates exceed the temperatures of their environs (Landsberg 1981; Oke 1982), promote species whose distribution is limited by cold temperatures, such as foreign species from southern climates (Sukopp *et al.* 1979). Lastly, cities are centers of trade and mobility, i.e. the junctions where traffic from different parts of the world meets and where (global) goods are transported to. Many species, e.g. the seeds of vascular plants, are transported intentionally (e.g. for cultivation) or accidentally together with goods or by traffic (von der Lippe & Kowarik 2007; 2008). As trade acts globally, species are spread globally. Thus, both native and foreign species tend to be introduced in urbanized areas (Kühn *et al.* 2004a), and foreign species accumulate more intensively in cities than in the countryside (Pyšek 1995; Sukopp 1998). While Barthlott *et al.* (1999) argued that the high urban species richness is an effect of sampling bias, this is not the case for Germany (Kühn *et al.* 2004a).

Given the increasing concentration of human activities in such diverse landscapes (Cincotta *et al.* 2000; Balmford *et al.* 2001; Sechrest *et al.* 2002), we should provide habitats to conserve species diversity not only in natural landscapes but also within urbanized areas (Rosenzweig 2003).

However, species richness is only one aspect of biodiversity. It is the aspect most often considered in ecological research (Purvis & Hector 2000) but it might not give enough details for a broad and effective conservation of species diversity.

The high urban species richness appears less diverse, when species inventories of cities worldwide are compared: The spread of species through global trade and traffic, and the similarity of urbanized environments worldwide homogenize urban floras all over the world, i.e. similar species are found in many cities worldwide (McKinney 2006). Thus, regionally, urban species richness increases total species richness but globally, urban species richness decreases total species richness (cf. Sax & Gaines 2003 for the global influence of exotic species).

3.2. Species Rarity

Beyond species richness, species frequency (or rarity) is a further aspect of biodiversity. Measures of species richness treat rare and common species equally. However, for conservation purposes, rare species are normally valued higher than common species. Therefore, it is important to consider species rarity in addition to species richness, especially in terms of species conservation.

Although a significant part of urban species richness is based on common native and alien species (Kühn & Klotz 2006), there are also urban habitats where rare species can persist (e.g. Brandes 1993). Some urban habitats act as analogues of natural habitats, such as buildings that act as rock surfaces or abandoned sand pits where heathland can develop (Eversham *et al.* 1996; Lenzin *et al.* 2007). Nevertheless, urbanization threatens especially rare native species while fostering already common species by destroying natural and semi-natural habitats (Kühn & Klotz 2006).

3.3. Functional Diversity

Another aspect of biodiversity, which is not covered by species richness, is functional diversity. Functional diversity is the diversity of species' traits and trait states; with traits being the characteristics of species, and states being the different categories of a categorical trait (e.g. insect-pollination is a state of the trait pollen vector). A species assemblage that consists of three wind-pollinated plants is functionally less diverse than a species assemblage of one wind-pollinated, one insect-pollinated and one self-pollinated plant, although their richness is the same (Fig. 12). This example illustrates that urban species assemblages might well be richer in species than non-urban species assemblages but at the same time functionally poorer, if only plants with certain trait states are able to persist in urbanized landscapes. The importance of functional diversity is shown by the following two examples: (i) With a sharp decrease of insect-pollinated plant species in the flora, many insects depending on nectar or pollen would go extinct and with them their predators (cf. Biesmeijer *et al.* 2006). Whole food webs would collapse (cf. Pauw 2007) and a significant amount of human food would disappear (Klein *et al.* 2007). (ii) If the specific leaf area (SLA = surface-to-weight ratio of leaves) of a significant percentage of plant species would decrease remarkably, the total flora's average degradation rate and mineralization cycles could slow down, with effects on nutrient supply and atmospheric gas cycles (cf. Wardle *et al.* 2004).

Because different traits have different functions, and because different functions fit to different environmental conditions, a species assemblage should be more stable if it is functionally diverse. Functional diversity is the flora's "toolbox" that enables it to react on a variety of environmental conditions and changes.

(a)   (b)

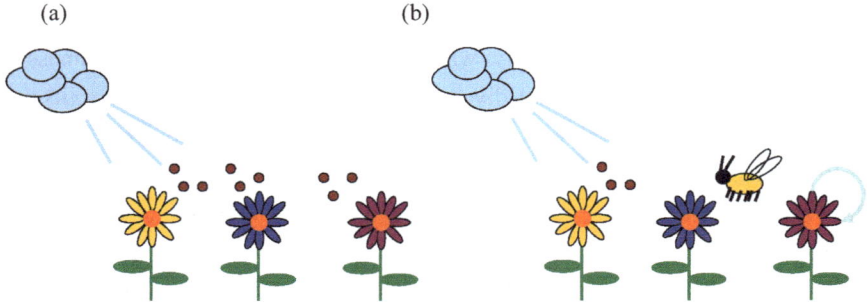

Figure I. 2 – An example of functional diversity
Two species assemblages, one (a) consisting of three wind-pollinated species, the other (b) consisting of one wind-pollinated, one insect-pollinated and one self-pollinated species. Both assemblages have the same species richness but (b) is functionally more diverse than (a).

In this study, Ellenberg values and floristic characterizations are used besides true species traits – traits have a genetic basis, Ellenberg values are habitat suitability indices based on expert guess, while floristic characterizations show, for example, the floristic origin of a species. Thus, strictly speaking, Ellenberg values and floristic characterizations are no traits. However, for simplicity, all species characteristics used in this study will be called traits from now on; their attributes will be called states.

3.4. Phylogenetic Diversity

Phylogenetic diversity is the diversity of evolutionary relationships (i.e. lineages) among species: A species assemblage that consists of three species belonging to the same family is phylogenetically less diverse than an assemblage of three species belonging to three different families, although their richness is the same (Magurran 2004; Fig. I3). Phylogenetic diversity is the basis for functional diversity: Many species' traits are heritable and conserved, meaning that species from different lineages often have a higher probability of having different trait states but species from the same lineage have a higher probability of sharing the same trait states (evolutionary niche conservatism; Harvey & Pagel 1991; Prinzing *et al.* 2001). However, if a trait is less conserved, species from the same lineage might differ in this trait but species from different lineages might share the trait state in a similar environment. Both cases illustrate that phylogenetic diversity influences the "toolbox" functional diversity.

(a)   (b)

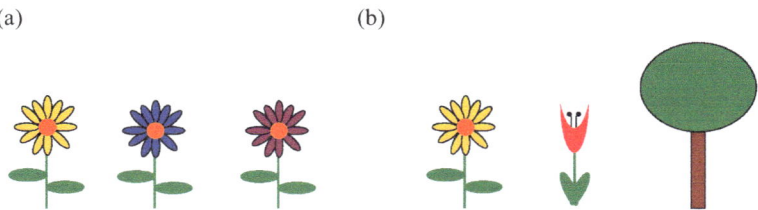

Figure I. 3 – An example of phylogenetic diversity
Two species assemblages, one (a) consisting of three species from the same family, the other (b) consisting of three species from three different families. Both assemblages have the same species richness but (b) is phylogenetically more diverse than (a).

4. Macroecology as an Analytical Framework

Classical ecological studies dealt (and still deal) with species assemblages on local or even smaller scales, while larger scales were often ignored (Gaston & Blackburn 2000). Local species assemblages however, are not solely influenced by local environmental conditions but also by regional and biogeographic parameters like large-scale climate (Hampe 2004), and local species pools depend on the available regional species pools (Zobel 1997). Moreover, biodiversity is threatened by forces acting globally, such as climate change, land-use change (including urbanization), biological invasions or nitrogen deposition (Sala *et al.* 2000). Therefore, it is necessary to consider large-scale patterns and processes to understand the structure, dynamics and characteristics of local species assemblages (Gaston & Blackburn 2000), which is the aim of the novel field of macroecology. "Macroecology is a way of studying relationships between organisms and their environment that involves characterizing and explaining statistical patterns of abundance, distribution, and diversity" (Brown 1995, p. 10). It "differs from most of recent and current ecology in its emphasis on statistical pattern analysis rather than experimental manipulation" (Brown 1995, p. 10). Macroecological analyses help to place the findings of small-scale analyses in a broader perspective. Thus, large-scale analyses need to complement small-scale analyses for an extensive understanding of human effects on biodiversity (Kühn *et al.* 2008).

Macroecological studies are not restricted to spatial studies; they can also include temporal analyses dealing with long time spans. Spatial analyses are able to reveal differences between different regions, different land-use types or else. Although they can partly replace temporal studies (space-for-time substitution), they reflect only one point in time. Contrastingly, temporal studies focus on processes and are able to show how a parameter of interest, such as the functional composition of a flora, changes in the course of a process, such as urbanization.

Macroecology is a well suited framework for urban ecological studies: Urbanization, suburban development and decentralisation created landscapes, which can neither be clearly defined as urban, nor as rural, because urban and rural land-use types intermingle. These landscapes illustrate that urbanization influences the environment across the administrative borders of cities. Therefore, its effects should be well detectable with macroecological techniques.

It is clear that in a strict sense, the term "urban flora" is restricted to the flora of typical urban-industrial habitats and does not apply to the flora of semi-natural remnants within a city (Wittig 2002). However, the surrounding urban matrix also influences remnants of semi-natural habitats, e.g. via fragmentation, nutrient inputs or disturbance. Thus, macroecological studies can reveal the influences of urban land use on the total flora of a study area, although study areas that include whole cities plus surroundings also contain habitats that are not typically urban-industrial.

According to the urban-rural landscapes created by urban sprawl, the terms "urbanized landscape", "urbanized region" or "urbanized area" are preferred here over "urban area", "urban region" or "urban landscape", at least when considering a large scale, and not the administrative borders of a city. However, this does not apply to terms such as "urban species richness" or "urban brownfield". Species richness cannot be urbanized in the sense of the word; the urban brownfield is a typical urban phenomenon.

In the present study, both spatial and temporal aspects of plant biodiversity in cities and urbanized areas are analyzed.

5. Study Outline

Cities harbor more species than their rural surroundings, at least over large enough scales (cf. Pautasso 2007). However, species richness does not necessarily cover all aspects of biodiversity, such as functional patterns, species rarity or phylogenetic relationships. Ignoring these relationships, our understanding of how species assemblages develop and change in a changing environment remains incomplete.

This study challenges the high vascular plant species richness of urbanized areas in Germany by providing insight into more detailed aspects of plant biodiversity. It shows differences between the functional composition of urban and rural floras by analyzing the frequency of plant trait states in urbanized, agricultural and semi-natural areas in Germany (Chapter I). It considers different spatial scales by analyzing the functional composition of urban and rural floras on a large scale with grid-cells sized c. 130 km² (Chapter I), and on a small scale comprising selected areas in the city of Halle (Saale) and its rural surroundings in Central Germany (Chapter II). Moreover, it shows how interactions of urban land use and species trait states influence plant species rarity (Chapter III). It considers the temporal dimension of biodiversity by analyzing changes in the functional composition of the flora of Halle over three centuries (Chapter IV). Additionally, the study shows how the phylogenetic diversity of plant assemblages differs between urbanized, agricultural and semi-natural areas in Germany (Chapter V). It relates phylogenetic diversity to species' trait states (Chapter V) and shows how species assemblages develop and change with a changing environment. All patterns are discussed in synthesis and put into the context of other urban ecological studies ("Synthesis and Conclusions").

For both nested study areas used in the analyses (Germany, and the city of Halle with its environs), urban-rural patterns of species richness were analyzed previously: Kühn *et al.* (2004a) showed for Germany that urbanized regions harbor more plant species than rural regions; Wania *et al.* (2006) showed that Halle is richer in plant species than its rural surroundings; and Knapp *et al.* (2008a) showed that protected areas in Halle are slightly poorer in vascular plant species richness than their counterparts in Halle's rural environs. Therefore, Germany and Halle are predestinated to analyze aspects of biodiversity that go beyond species richness.

Detailed questions analyzed and discussed are:

Chapter I: Do urbanized environments, which differ from agricultural and semi-natural environments in many respects, favor plant species with other trait states than rural environments? Which characteristics of urbanized and rural environments cause differences in the functional composition of floras?

Chapter II: Do differences in the functional composition of urban and rural floras that exist on large spatial scales also hold for small-scale urbanization gradients?

Chapter III: Do urban land use and plant species' trait states influence plant species rarity? Do plant species with specific trait states differ in rarity when having a high or low affinity to urban land use?

Chapter IV: Does urbanization, which deeply changes the characteristics of a landscape, cause temporal changes in the functional composition of floras? Do spatial differences in the functional composition of today's urban and rural floras reflect temporal changes of functional composition caused by urbanization?

Chapter V: Do floras from urbanized and rural areas differ in phylogenetic diversity? Do differences in phylogenetic diversity between urbanized and rural areas reflect differences in species richness? Does the phylogenetic diversity of groups of plant species sharing specific trait states change systematically between urbanized and rural areas?

Chapter I – Urbanization Causes Shifts of Species' Trait State Frequencies – a Large Scale Analysis

1. Introduction

Cities differ from rural landscapes in many ways: Human densities peak in urbanized areas; pollutants, energy and nutrients concentrate there (Sukopp 1998); exotic species are more frequent and land use is more heterogeneous (Niemelä 1999; McKinney 2002; Kühn *et al.* 2004a); annual average air temperature is 0.5 to 1.5°C higher than in the non-urban surroundings and air moisture is reduced, at least in temperate and boreal zones (Sukopp 1998). In contrast to cities, many agricultural landscapes are homogeneous over large areas. They are often subjected to a highly industrialized agriculture, characterized by high pesticide and fertilizer input and water management aiming at the maintenance of favorable soil moisture conditions. Forested and semi-natural landscapes are often nutrient poor, like forests on siliceous rock or heathland, because many nutrient rich habitats were transformed to agricultural or urban habitats (cf. Pressey 1994).

Differences in land use lead to differences in species composition since functional traits (such as pollination) have different states (i.e. different classes of categorical traits, e.g. wind-pollination, insect-pollination, self-pollination) which respond differentially to environmental gradients and therefore show distinct biogeographic patterns (e.g. Kühn et al. 2006). In other words: Different environments filter for species with different trait states (Zobel 1997). An example is the study of Wittig and Durwen (1982) comparing the spectra of environmental indicator-values (Ellenberg et al. 2001) of spontaneous floras in four cities in the West of Germany with the floras of the cities' rural surroundings, showing a greater proportion of high indicator-values for e.g. light, temperature and nitrogen in the cities. Similar results were obtained for the Central German city of Halle (Klotz 1989) and the Czech city of Plzeň (Chocholoušková & Pyšek 2003). Therefore, shifts in land use, e.g. increasing urbanization accompanied by an increase in temperature (Landsberg 1981; Oke 1982; Sukopp 1998), might lead to shifts in trait state frequency and, in extreme cases, to the loss of plants with certain trait states (Díaz & Cabido 1997). Tamis et al. (2005) showed that recent changes in the frequency of occurrence of vascular plant species across the Netherlands are at least partly related to both urbanization and climate change. They did, however, not consider shifts in trait state frequency. If these shifts in trait spectra indeed occur, they might affect ecosystem functioning: Increased leaf dry matter content for example might decrease litter decomposability (Kazakou et al. 2006).

Today's differences in the trait state composition of urban and rural floras might point to potential future shifts with further urbanization. We compared the proportions of several trait states of vascular plants in urbanized, agricultural and semi-natural areas in Germany. We ask which trait states enable a plant to cope with the specifics of urban environments, e.g. the urban climate (Sukopp 1998), irregular disturbance, and spatial and temporal heterogeneity (Niemelä 1999). We chose traits that we expect to respond to these urban conditions: Leaf traits [leaf anatomy, leaf persistence, specific leaf area (SLA), leaf dry matter content (LDMC)] should respond to climate, because gas exchange and water storage make leaves key organs regarding the adaptation to air temperature and moisture (cf. Wright et al. 2005). Further, we chose type of reproduction, life span, and life form, as traits related to persistence and regeneration after disturbance (which is true for SLA and LDMC as well; Lavorel & Garnier 2002; Wittig 2002; Sudnik-Wójcikowska & Galera 2005). Both spatial and temporal heterogeneity call for the ability of plants to disperse in space, therefore we included dispersal type as another trait. Poschlod & Bonn (1998) already claimed shifts in dispersal processes in man-made landscapes, especially after land-use intensification which may cause the decrease or increase of species frequency (Römermann et al. 2008). Last, we chose pollination type and UV-reflection of flowers; both are related to a plant's fecundity and reflect the suitability of the environment for pollinating insects. We discuss possible urban and rural filters and consequences of shifts in trait expression.

2. Materials and Methods

2.1. Data Sources

Data on species' traits originate from BiolFlor, a database on biological and ecological traits of the German flora (Klotz *et al.* 2002; http://www.ufz.de/biolflor; Kühn *et al.* 2004b) and from LEDA, a database on life-history traits of the Northwest European flora (Kleyer *et al.* 2008; http://www.leda-traitbase.org; see Table A1 in the appendix for a complete over-view and description of traits and trait states).

Plant species occurrences originate from the database on the German flora (FLORKART, http://www.floraweb.de), maintained by the German Center for Phytodiversity at the Federal Agency for Nature Conservation (Bundesamt für Naturschutz BfN). In FLORKART, Germany is divided into grid-cells of 10 minutes longitude × 6 minutes latitude (corresponding to *c.* 12 × 11 km or 130 km²). The database contains more than 14 million records of plant occurrences, acquired by thousands of volunteers. We did not use cultivated occurrences and only referred to the spontaneous flora. This means that occurrences of cultivated alien species that escaped from cultivation and form self-sustaining populations were mapped when occurring spontaneously. This applies e.g. to *Robinia pseudoacacia* L. or *Ailanthus altissima* (Mill.) Swingle when growing spontaneously on e.g. railway sites. Occurrences that were apparently planted, e.g. in any kind of garden, urban park or on cemeteries, were not included in the analyses. All plant occurrences mapped by the volunteers were controlled for plausibility by specialists in floristic recording centers (see e.g. http://www.biologie.uni-regensburg.de/Botanik/Florkart/dbblber.htm). However, mapping intensity varies among grid-cells. Therefore, we only used grid-cells with at least 45 of 50 con-trol species. These are the 45 most frequent species of the German flora according to Krause (1998) plus five generalists considered by the volunteers to be difficult to determine (Kühn *et al.* 2004a; Kühn *et al.* 2006). 136 out of 2995 grid-cells were excluded due to an insufficient number of control species.

Land-use data per grid-cell are based on Corine Land Cover data that are derived from satellite remote sensing images (Statistisches Bundesamt 1997; http://www.corine.dfd.dlr.de/intro_en.html). Corine data differentiate between artificial (i.e. urban), agricultural, and forested/semi-natural land use, wetlands and water bodies. These land-use classes again are subdivided; the artificial land-use class for example includes built-up residential, industrial, commercial and transport area, mines, dumps, and artificial non-agricultural vegetated areas, i.e. urban green space. However, we only used the main classes: We classified grid-cells with more than 33% of urban land use as urbanized grid-cells (n=59) (Kühn & Klotz 2006) and split the remaining grid-cells into agricultural ones with more than 50% agriculture (n=1365) and semi-natural ones with more than 50% forests or semi-natural land use (n=312; Fig. A2 in the appendix). Grid-cells not meeting the selection criteria were omitted (n=1259).

To account for effects of other environmental parameters than land use on the trait state proportions, we used co-variables calculated per grid-cell and known to act on species diversity (Kühn *et al.* 2003). Data on climate [mean annual, mean July and mean January temperature, mean difference between July and January temperature (all 1961-1990), mean annual precipitation (1951-1980), mean wind speed] were provided by the "Deutscher Wetterdienst Department Klima und Umwelt"; data on topography (mean altitude above sea level) were provided by ESRI (ARCDeutschland 500 dataset, 1: 500,000); data on soils (number of soil types, number of soil patches), and geology (number of geological types, number of geological patches) are based on the German soil survey map (Bundesanstalt für Geowissenschaften und Rohstoffe 1995) and the Geological survey map (Bundesanstalt für Geowissenschaften und Rohstoffe 1993) provided by the German Federal Agency for Nature Conservation. For an overview of differences in environmental parameters between urbanized and rural grid-cells in Germany see Fig. A1 in the appendix.

2.2. Data Analyses

2.2.1. Log-Ratios of Proportions

We merged the matrices on species per grid-cell and trait state per species (by matrix multiplication) to a matrix on trait state frequency per grid-cell, from which we calculated the trait state proportions (for numbers of species analyzed per trait state see Table A2 in the appendix). Because the proportions add up to 100%, they depend on each other. To break this unit sum constraint, we used log-ratios of proportions (e.g. Aitchison 1982; Billheimer *et al.* 2001; Kühn *et al.* 2006). The log-ratio of two trait states a and b is log (a/b). For traits with more than two states the denominator should always be the same, without relevance which trait this is.

Zero values can neither be log-transformed nor used in the denominator. Therefore, we replaced each zero with the proportion one trait state would have if expressed by only one out of all species of a grid-cell, and reduced the respective non-zero values by a corresponding amount (Fry *et al.* 2000; Martin-Fernandez *et al.* 2000). Each log-ratio was used separately in the further analyses.

SLA and LDMC are the only continuous traits in our analysis, i.e. they were used directly as responses in the linear models without preceding log-transformation.

2.2.2. Linear Models

To minimize the effects of climate, topography, soils and geology on the trait state composition, we explained each log-ratio in a multiple linear regression with the corresponding parameters. We included selected two-way interactions and reduced each model via backward selection until achieving its minimal adequate version (model selection by AIC; Mac Nally 2000). We calculated the mean of the minimal adequate models' residuals (i.e. the variation not explained by climate, topography, soils and geology) per grid-cell type. Since there are more agricultural (n=1365) and semi-natural (n=312) than urbanized (n=59) grid-cells, we resampled the former two separately by calculating the mean of 59 randomly chosen grid-cells 999 times. We tested for significant differences between the mean residuals of urbanized and agricultural and between urbanized and semi-natural grid-cells with the z-statistic (comparison of one mean value to a distribution of mean values).

We also calculated differences between urbanized and non-urbanized grid-cells with an alternative method, by including the three land-use types as a categorical predictor in a linear model together with the environmental parameters on climate, topography, soils and geology, and explaining each log-ratio with these predictors (Knapp *et al.* 2008b). This approach yielded the same results as the resampling-approach and is not presented here.

Choosing environmental variables to minimize non-land-use effects on log-ratios is problematic, because we might miss important variables. To corroborate the results of our first analysis, we additionally explained the log-ratios using linear mixed effect models that allow for random and fixed effects: We assigned the urbanized grid-cells to six regions that are reasonably homogeneous with respect to biogeography (Fig. 1.1). Within each region, we selected as many agricultural and semi-natural grid-cells as there were urbanized grid-cells to account for the differences in sample size of grid-cell types.

We explained the log-ratios (and SLA and LDMC) with the regions as random effects and land use as fixed effect. On this, we performed a variance components analysis.

We performed all analyses with the open source software R, Version 2.3.1 (R Development Core Team 2006), calculating the linear mixed-effect models with the R-function lme from the package nlme (Pinheiro *et al.* 2006), and variance components analysis with the R-function varcomp from the package ape (Paradis *et al.* 2006).

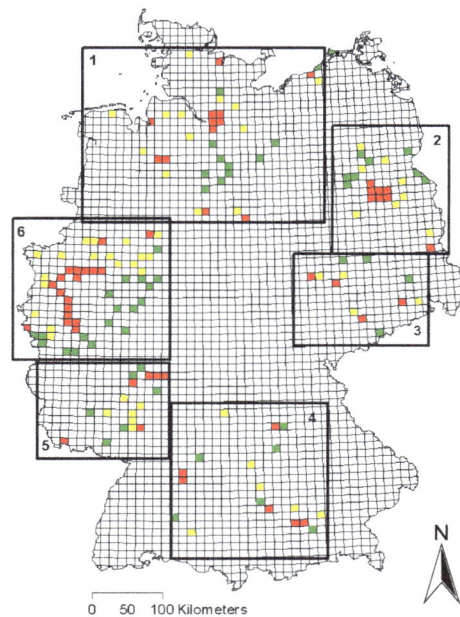

Figure 1. 1 – Six regions in Germany selected for comparison of effects of biogeography and land use on the functional composition of plant species assemblages
(1) Northern Germany; (2) Berlin and Brandenburg; (3) Saxony and Saxony-Anhalt; (4) Southern Germany; (5) Rhine-Main region; (6) Rhine-Ruhr region. Red: urbanized grid-cells; yellow: agricultural grid-cells; green: semi-natural grid-cells. Figure taken from Preslia 80, 375-388 (2008), reprinted with permission of the Czech Botanical Society

3. Results

The models correcting for climate, topography, soils and geology were all significant and explained between 9% and 71% of the log-ratios (Table 1.1). Most differences in trait state composition between urbanized and non-urbanized grid-cells were significant (Fig. 1.2, Table 1.2): Plants dispersed by animals, humans or water had increased proportions in urbanized grid-cells at the expense of plants dispersed by wind, which were relatively more frequent in agricultural and semi-natural grid-cells (Fig. 1.2a, b). Proportions of plants with hygromor-phic leaves were decreased in urbanized grid-cells in favor of plants with meso-morphic, scleromorphic or succulent leaves. (Fig. 1.2c, d). LDMC was lower in urbanized than in both types of non-urbanized grid-cells (Fig. 1.2e). Plants with overwintering green leaves were more frequent in urbanized than in agricultural and semi-natural grid-cells (Fig. 1.2f). There were relatively more therophytes in urbanized than in non-urbanized grid-cells but accordingly less chamaephytes, geophytes, hemicryptophytes and phanerophytes in proportion to therophytes in the urbanized grid-cells (Fig. 1.2g, h).

Plants in urbanized grid-cells were more often annual or biennial (Fig. 1.2i) than plants in agricultural and semi-natural grid-cells. Urbanized grid-cells had more wind-pollinated plants but less insect- and self-pollinated plants than agri-cultural grid-cells but showed no differences to semi-natural grid-cells (Fig. 1.2j, k). Plants in urbanized grid-cells had a higher SLA than plants in semi-natural grid-cells but showed no differences to plants in agricultural grid-cells (Fig. 1.2l). Plants in urbanized grid-cells reproduced more often by seeds than plants in non-urban grid-cells (Fig. 1.2m). Lastly, there were more plants with UV-reflecting flowers in urbanized than in agricultural and semi-natural grid-cells (Fig. 1.2n).

The linear mixed effect models correcting for biogeographic effects mainly corroborated these results. Land use explained more variance than the bio-geographic differences between the six regions, throughout all tested trait state ratios (Table 1.3).

Table 1. 1 – Variation of trait state ratios in urbanized, agricultural, and semi-natural grid-cells in Germany explained by climate, topography, soils, and geology in multiple linear models
R^2 shows the variation and is adjusted for the number of predictors. Asterisks mark significant P-values: *** = $P \leq 0.001$.

Trait	Trait state ratios	R^2
Dispersal type	Dysochorous/Anemochorous	0.45***
	Endozoochorous/Anemochorous	0.53***
	Epizoochorous/Anemochorous	0.56***
	Hemerochorous/Anemochorous	0.56***
	Hydrochorous/Anemochorous	0.63***
Leaf anatomy	Helomorphic/Hygromorphic	0.51***
	Hydromorphic/Hygromorphic	0.49***
	Mesomorphic/Hygromorphic	0.40***
	Scleromorphic/Hygromorphic	0.37***
	Succulent/Hygromorphic	0.22***
LDMC	-	0.18***
Leaf persistence	Evergreen / overwintering green	0.67***
	Spring green / overwintering green	0.29***
	Summer green / overwintering green	0.65***
Life form	Chamaephyte/Therophyte	0.71***
	Geophyte/Therophyte	0.59***
	Hemicryptophyte/Therophyte	0.70***
	Hydrophyte/Therophyte	0.30***
	Phanerophyte/Therophyte	0.45***
Life span	Annual/pluriennial	0.70***
	Biennial/pluriennial	0.46***
Pollen vector	Insects/wind	0.52***
	Selfing/wind	0.41***
SLA	-	0.24***
Type of reproduction	Generatively only/vegetatively only	0.48***
	Generatively & vegettively/vegetatively only	0.66***
UV-reflection of flowers	No/yes	0.09***

Table 1. 2 – Differences between the functional composition of the flora in urbanized, agricultural, and semi-natural grid-cells in Germany

u = urbanized; a = agricultural; sn = semi-natural. *P*-values: 0.05< *P* ≤ 0.1+, *P* ≤ 0.05*, *P* ≤ 0.01**, *P* ≤ 0.001***, for non significant differences equal values are assumed.

Trait	Trait state	Urbanized – agricultural	Urbanized – semi-natural
Dispersal type	Dysochorous/Anemochorous	u < a +	u < sn*
	Endozoochorous/Anemochorous	u > a**	u > sn**
	Epizoochorous/Anemochorous	u > a***	u > sn***
	Hemerochorous/Anemochorous	u > a***	u > sn***
	Hydrochorous/Anemochorous	u > a***	u > sn***
Leaf anatomy	Helomorphic/Hygromorphic	u > a**	u = sn
	Hydromorphic/Hygromorphic	u = a	u > sn**
	Mesomorphic/Hygromorphic	u > a***	u > sn***
	Scleromorphic/Hygromorphic	u > a***	u > sn***
	Succulent/Hygromorphic	u > a***	u > sn***
LDMC	-	u < a***	u < sn***
Leaf persistence	Evergreen / overwintering green	u < a***	u < sn***
	Spring green / overwintering green	u = a	u < sn*
	Summer green / overwintering green	u < a***	u = sn
Life form	Chamaephyte/Therophyte	u < a***	u < sn***
	Geophyte/Therophyte	u < a***	u < sn***
	Hemicryptophyte/Therophyte	u < a***	u < sn***
	Hydrophyte/Therophyte	u < a***	u < sn +
	Phanerophyte/Therophyte	u < a***	u < sn***
Life span	Annual/pluriennial	u > a***	u > sn***
	Biennial/pluriennial	u > a***	u > sn***
Pollen vector	Insects/wind	u < a*	u = sn
	Selfing/wind	u < a +	u = sn
SLA	-	u = a	u > sn**
Type of repro-duction	Generatively only/vegetatively only	u > a***	u > sn**
	Generatively & vegettively/vegetatively	u < a*	u = sn
UV-reflection of	No/yes	u < a**	u < sn***

Table 1. 3 – Variation [%] of the flora's functional composition explained in linear mixed effect models by land use and differences in biogeography between six regions in Germany that are relatively homogeneous with respect to biogeography. See below for further explanation.

Trait	Trait state ratio	Land use	Biogeography	Int_u	Int_a	Int_{sn}
Dispersal type	Dysochorous/Anemochorous	80	20	-0.67	-0.62**	-0.67 n.s.
	Endozoochorous/Anemochorous	63	37	0.05	0.04 n.s.	0.01*
	Epizoochorous/Anemochorous	78	22	0.60	0.51***	0.45***
	Hemerochorous/Anemochorous	88	12	0.37	0.32**	0.24***
	Hydrochorous/Anemochorous	53	47	-0.45	-0.47 n.s.	-0.56***
Leaf anatomy	Helomorphic/Hygromorphic	62	38	0.29	0.25 n.s.	0.16***
	Hydromorphic/Hygromorphic	83	17	-1.25	-1.30 n.s.	-1.65***
	Mesomorphic/Hygromorphic	94	6	1.58	1.44***	1.41***
	Scleromorphic/Hygromorphic	77	23	0.57	0.35***	0.34***
	Succulent/Hygromorphic	96	4	-2.37	-2.63***	-2.72***
LDMC		68	32	211.62	218.70***	223.33***
Leaf persistence	Evergreen / overwintering green	89	11	0.82	0.93***	1.07***
	Spring green / overwintering green	91	9	-1.78	-1.74 n.s.	-1.94*
	Summer green / overwintering green	81	19	1.51	1.55 n.s.	1.59***
Life form	Chamaephyte/Therophyte	89	11	-1.99	-1.88***	-1.60***
	Geophyte/Therophyte	89	11	-0.84	-0.68***	-0.58***
	Hemicryptophyte/Therophyte	87	13	0.70	0.86***	0.97***
	Hydrophyte/Therophyte	78	22	-1.74	-1.53**	-1.75 n.s.
	Phanerophyte/Therophyte	84	16	-0.72	-0.54***	-0.42***
Life span	Annual/pluriennial	89	11	16.33	16.22***	16.13***
	Biennial/pluriennial	89	11	15.17	15.13**	15.11***
Pollen vector	Insects/wind	55	45	1.02	1.03 n.s.	1.04 n.s.
	Selfing/wind	59	41	0.84	0.85 n.s.	0.85 n.s.
SLA		75	25	26.99	26.94 n.s.	26.74*
Reproduction	Generatively only/vegetatively only	61	39	2.27	2.16**	2.17**
	Generatively & vegetatively/veg. only	61	39	1.93	1.95 n.s.	2.02***
UV-reflection	No/yes	96	4	0.45	0.46 n.s.	0.48**

Further explanation for table 1. 3: "Land use" and "Biogeography" show the percentage of variation explained by the respective parameters. Model intercepts for urbanized (Int_u), agricultural (Int_a) and semi-natural (Int_{sn}) grid-cells show whether there are significant differences between urbanized and agricultural (see Int_a) or between urbanized and semi-natural grid-cells (see Int_{sn}). Levels of significance are indicated as follows: n.s. = not significant, $* = P \leq 0.05$, $** = P \leq 0.01$, $*** = P \leq 0.001$.

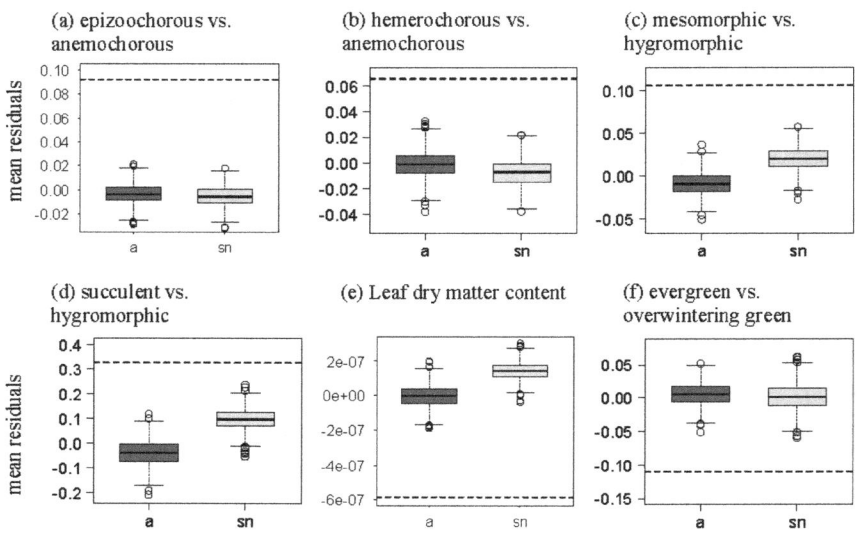

Figure 1. 2 – Frequency of trait states in urbanized, agricultural and semi-natural grid-cells in Germany, corrected for specific environmental co-variables.

Shown are selected results: (a) epizoochorous vs. anemochorous; (b) hemerochorous vs. anemochorous; (c) mesomorphic vs. hygromorphic; (d) succulent vs. hygromorphic; (e) leaf dry matter content; (f) evergreen vs. overwintering green; (g) geophytes vs. therophytes; (h) phanerophytes vs. therophytes; (i) annuals vs. pluriennials; (j) insect-pollinated vs. wind-pollinated; (k) self-pollinated vs. wind-pollinated; (l) specific leaf area; (m) generative reproduction vs. vegetative reproduction; (n) flowers not reflecting UV vs. UV-reflecting flowers. Boxplots represent median (line), 25-75 % quartiles (boxes), ranges (whiskers) and extreme values (circles). Dark grey = agricultural grid-cells; light grey = semi-natural grid-cells; dashed line = urbanized grid-cells. Values for agricultural and semi-natural grid-cells are based on resampling. Shown are residuals (see Materials and Methods section of this chapter for details). P-values for differences be-

tween urbanized and agricultural/ urbanized and semi-natural grid-cells are shown in
Table 1.2.

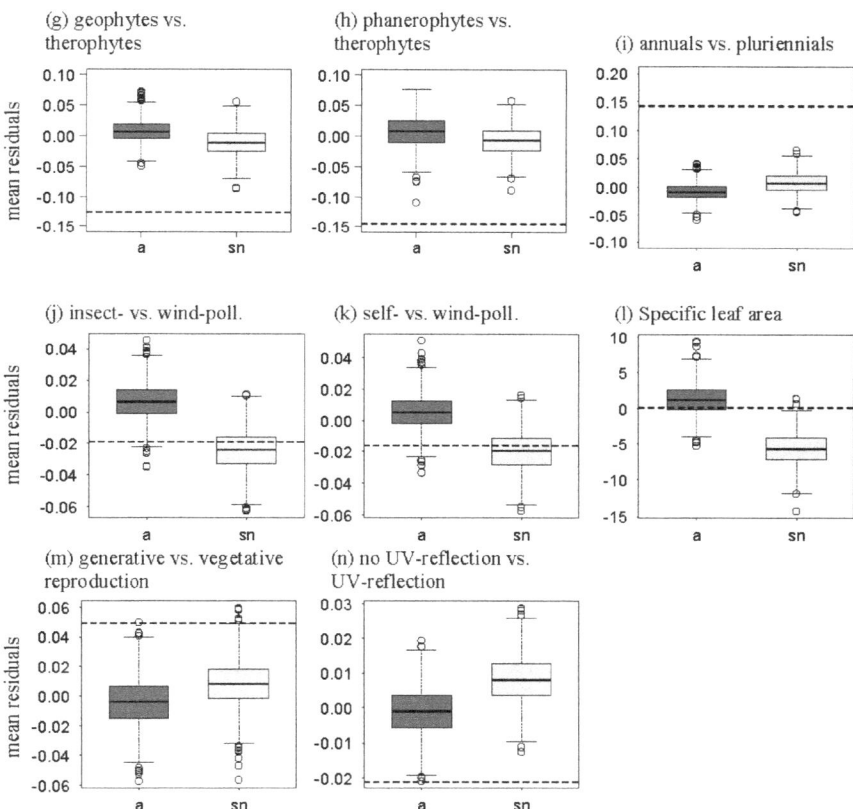

Figure 1.2. – continued

4. Discussion

The urban environment clearly favors plants with other trait states than agricultural or semi-natural environments. The trait state patterns we found are likely
due to typical urban filters: First, the urban heat island (Landsberg 1981; Oke

1982) favors plants either able to cope with drought, e.g. plants with succulent or scleromorphic leaves, or to avoid drought, e.g. annuals that finish their life cycle in a temporal niche like springtime, when temperatures and drought stress are still low (Wittig 2002). High temperatures also promote plants with overwintering green leaves by decreasing the risk of frost, as already shown by Wittig & Ou (1993) for the *Hordeetum murini*, an association that is very typical for Central European cities. Furthermore, low air moisture promotes wind-pollination by increasing the probability of pollen to reach receptive surfaces (Culley *et al.* 2002). Secondly, the intensive and irregular disturbances in urbanized areas favor annuals and biennials (Kleyer 1999), leaves with high specific leaf area, low leaf dry matter content (Díaz *et al.* 1999), and plants with overwintering green leaves. The latter are often short-lived and use a temporal niche: In winter, disturbance in gardens, parks and cemeteries reaches a minimum. Thirdly, the spatial and temporal heterogeneity of cities should promote plants with high dispersal abilities. Although all dispersal types we compared potentially enable long-distance seed dispersal (Knevel *et al.* 2005), our results suggest that wind is less adequate for dispersal in urbanized areas: Wind gets channeled in streets and often follows the increasing temperatures towards the city center, thus seeds should end up more often on sealed surfaces. Moreover, calms are more frequent in than outside cities (Kuttler 1993) and seeds do not reach potential habitats in the lee of houses or walls. In contrast to wind-dispersal, animal-dispersal (endo- and epizoochory) seems to work as well in urbanized areas as dispersal by humans does. On the one hand, birds, cats, dogs, and some wild mammals like foxes (Gloor *et al.* 2001) are potential dispersers. However, birds are mainly relevant for the dispersal of fleshy fruits and hardly cover other types of seeds (Kollmann 1994). On the other hand, animal-dispersal and human-dispersal might overlap, with species with adhesive dispersal using humans or even cars as vehicles instead of animals (Hodkinson & Thompson 1997; von der Lippe & Kowarik 2007; 2008).

Fourthly, the high percentage of sealed surfaces increases the amount of surface runoff (Wessolek & Renger 1998), which in turn should be beneficial for plant species that disperse by water: Rain water that percolates immediately into the soil after reaching the earth surface cannot transport seeds; water running down a street can transport seeds over longer distances, e.g. to the next roadside ditch. Irrigation of urban green spaces might as well be advantageous for hydro-chorous species.

Finally, cities have a high proportion of unstable habitats (e.g. urban brown-fields) that favor annuals and biennials, reproduction by seeds and therophytes (Brandes & Oppermann 1995; Wittig 2002; Sudnik-Wójcikowska & Galera 2005). Note that therophytes are annual and reproduce by seeds (Table A1), thus, the results confirm each other since the trait states are correlated and partly de-pend on the same environmental factors. The same is true for specific leaf area and low leaf dry matter content, which are negatively correlated (Roche *et al.* 2004). Higher SLA and higher proportions of plants with scleromorphic leaves in urbanized areas seem contradictory but again point to the high heterogeneity of urbanized areas (Niemelä 1999) with dry, warm habitats like urban brown-fields that support scleromorphic leaves, and nutrient rich watered habitats like urban parks, gardens and cemeteries (Sukopp 1998) that support high SLA (cf. Wright *et al.* 2005).The pattern for evergreen species seems contradictory as well: The high urban temperatures should not only promote plants with overwin-tering green leaves but also evergreen plants. However, the latter are more fre-quent in non-urban than in urbanized grid-cells. Thus, temperature is not the only restricting factor for evergreen species but land use is the main driving force: Evergreen species are normally long-lived and thus sensitive to distur-bance.

The rural filters oppose the urban filters: Temperatures are lower, distur-bances are more regular (in agricultural habitats) or less frequent (in semi-natural habitats), land use is more homogeneous (Lososová *et al.* 2006). Addi-tionally, rural environments seem to be more suitable for insects than urban envi-ronments, due to less pollution and different land-use structure (e.g. less built-up area) and consequently favor insect-pollinated plants over wind-pollinated plants (Lososová *et al.* 2006). It seems contradictory having more wind-pollinated but less wind-dispersed species in urbanized grid-cells but pollinating insects might be more sensitive to urban land use than seed-dispersing animals, and animal-dispersed species might also be dispersed by humans (see above). A sensitivity of insects to urban land use can also explain the higher frequency of plants with UV-reflecting flowers in urbanized grid-cells: In BiolFlor, UV-reflection is mainly documented for insect-pollinated plants, which have to struggle harder in urbanized than in rural areas to attract their visitors. Nevertheless, we cannot tell from our data whether a low proportion of insect-pollinated species decreases pollinator richness or vice versa. It might be a parallel response to urbanization (Biesmeijer *et al.* 2006).

It seems surprising that self-pollinated species are less frequent in urbanized than in rural areas, although many urban habitats are quite young. Newly created habitats should be first invaded by plants that are independent of insects or other pollinating animals, since if pollinator availability is low, a successful reproduction is only possible by wind or self-pollination (Culley *et al.* 2002; Düring 2004). However, selfers are only more frequent in agricultural than in urbanized grid-cells but show no difference between semi-natural and urbanized grid-cells. Agricultural areas are tilled and harvested often several times a year. Therefore, many agricultural habitats are even younger than urban brownfields or industrial habitats. Furthermore, we did not include abundance data in our analysis, because these are not available for the total German flora. Including abundance data might clarify the pattern for self-pollinated species: Wittig (2002) showed that of the 20 plant species which are most common in the flora of Central European cities, 70% are self-pollinated.

Germany covers a range of biogeographic regions from the Alps in the South to the coasts in the North. There are more cities in the North and West of Germany but less in the South (Fig. A2); most cities are situated on rivers and below 300m a.s.l. (Kühn & Klotz 2006). Thus they have a biogeographically biased distribution (Kühn *et al.* 2004a). Therefore, the trait state patterns might not reflect differences between urban and rural land use but biogeographic gradients. Although we accounted for several parameters on climate, topography, soils and geology, there are of course more environmental parameters that might influence trait state patterns, such as variation in altitude, sunshine duration or length of vegetation period. To account for all these biogeographic effects, we applied the linear mixed effect models. Nevertheless, land use explains even more variation than differences in biogeography (Table 1.3).

Our results might be influenced by phylogenetic relatedness of species and spatial autocorrelation. Both can alter parameter estimates of linear models (Kühn 2007; Tremlová & Münzbergová 2007). Though we are aware of this, we neither corrected for phylogeny nor for spatial autocorrelation. We think that our results are yet reliable: Firstly, including or excluding phylogenetic relatedness produced similar results for most traits in an urban-rural comparison of plant trait patterns in the Czech Republic (Lososová *et al.* 2006). Secondly, analyses without phylogenetic correction are less problematic when dealing with large rather than small species groups (Tremlová & Münzbergová 2007). Besides, we are not aware of any method suitable to account for spatial as well as phyloge-netic autocorrelation.

Our study clearly shows that on a coarse spatial scale shifts in land use can change the trait state composition of plant assemblages. This finding is remark-able, given the fact that grid-cells are rather heterogeneous – there is still 66% of non-urban land use in a grid-cell with 34% of urban land use. However, modern cities are not restricted to a few square kilometers bordered by city walls. They rather spread in the surroundings where they mingle with rural land use, creating urbanized landscapes. Given this spatial heterogeneity it can therefore be ex-pected that, in addition to the effects of urban land use on coarse spatial scales, there are additional effects of urbanization on smaller spatial scales. There is, however, some evidence that the positive relation between urban land use and species richness (e.g. Hope *et al.* 2003; Araújo 2003; Kühn *et al.* 2004a) is espe-cially strong at coarse scales (Pautasso 2007).

In conclusion, our study shows that shifts in land use can change the trait state composition of plant assemblages. Strong urbanization might consequently homogenize our flora with respect to trait state frequency.

Chapter II – Does Urbanization Cause Shifts of Species' Trait State Frequencies? – A Small Scale Analysis

1. Introduction

Urbanization is a global phenomenon that acts on large scales (cf. Vitousek *et al.* 1997). Therefore, large-scale analyses are well suited to reveal influences of urbanization on the composition of species assemblages. However, nature conservation often requires local action and nature conservation areas rarely exceed a regional level. Moreover, both large-scale climate and small-scale environmental conditions influence species ranges and therefore species occurrences (Korneck *et al.* 1998; Hampe 2004).

Results from large scales cannot simply be transferred to small scales, because patterns of diversity often differ between different scales (Kühn & Klotz 2007). Sax and Gaines (2003) for example, demonstrated that the introduction of exotic species often increases species numbers at local and regional scales but decreases species numbers globally. Pautasso (2007) showed that the correlation of human presence and species richness is positive at large scales but levels off or even turns negative the smaller the observed scale gets. Consequently, "translating" large-scale biodiversity patterns into small-scale biodiversity conservation asks for small-scale analyses complementing large-scale analyses.

The analyses presented in Chapter I showed how urbanization changes the functional composition of Central European floras, but the results only provide assumptions for species conservation activities in urbanized areas. Analyses of the flora's functional composition along small-scale urbanization gradients should provide more valuable information for species conservation in cities by showing whether specific habitats support species assemblages with a more "non-urban" functional composition and thus enable a high functional diversity.

We repeated the analyses presented in Chapter I on a smaller scale by choosing (i) protected areas and (ii) randomly selected plots sized 0.06 km² each in the city of Halle (Central Germany) and its rural surroundings (the former districts of Saalkreis and Mansfelder Land, now part of the new districts of Saalekreis and Mansfeld-Südharz). The protected areas in Halle are semi-natural elements within the urban matrix. The 0.06 km²-plots are scattered over the city and should represent a cross-section of the city's habitats, also including more semi-natural elements like urban parks or gardens. Consequently, both should provide habitats for species assemblages more typical of semi-natural habitats "in the midst of human enterprise" (Rosenzweig 2003).

2. Materials and Methods

2.1. Study Area

The city of Halle is situated in Central Germany, south-east of the Harz Mountains (city center: 11° 58' 19" E, 51° 28' 59" N; Fig.2.1). It covers an area of 135 km². With a mean annual temperature of 9°C (range of mean monthly temperature *c.* 0-19°C) and an annual precipitation of 480 mm with a peak in summer, the climate is subcontinental and relatively dry, at least in the Central European context (Müller-Westermeier *et al.* 1999; 2001). The low precipitation is caused by the rain shadow of the Harz Mountains. The climate enables the cultivation of wine, apples, cherries and apricots west of Halle. Without anthropogenic influence, forests of sessile oaks (*Quercus petraea* Liebl.), hornbeam (*Carpinus betulus* L.), and basswood (*Tilia cordata* Mill.) would be the main zonal vegetation (Institut für Länderkunde Leipzig 2003).

The river Saale flows through the study area, at an altitude of approximately 70 m a.s.l. Within the city of Halle, the river divides in several arms, forming islands and floodplain forests. South of Halle, the rivers Saale and Elster together form a larger floodplain area. The Saale valley in the northern part of the city and in the district of Saalkreis, north-west of the city, is characterized by porphyric rocks that border the valley. The south-western part of Halle is built on

Triassic and Tertiary bedrock. The north-eastern and the south-western parts of the city and its surroundings are divided by a fault line that runs directly through the city center giving rise to a salt spring from Late Permian (Wagenbreth & Steiner 1982). In the eastern part of the city, older bedrocks are nearly completely overlaid by quaternary bedrock.

During the last ice-age, the region was located at the southern edge of the glaciers covering Northern Europe where loess accumulated (Lang 1994). Hence, soils in the study area are mainly Chernozems (according to FAO classification) which are highly suitable for agriculture (Ministerium für Raumordnung 1996). Therefore, the rural surroundings of Halle are dominated by agricultural land use.

2.2. Data Sources

We chose protected areas > 0.06 km^2 in the city of Halle (n=14) and the surrounding rural district of Saalkreis (n=13) (Fig. 2.1; digital maps provided by the Environmental State Agency Saxony Anhalt and the city administration of Halle; Stadt Halle 2003a). We further chose twenty randomly selected 0.06 km^2-plots in Halle and the districts of Saalkreis and Mansfelder Land, respectively, from Wania et al. (2006; Fig. 2.2).

Data on plant species occurrences in the protected areas originate from regional species lists (Buschendorf & Klotz 1996; Landesamt für Umweltschutz Sachsen-Anhalt 2005). To evaluate the mapping intensity of protected areas, we used the semi-logarithmic species-area-curve (Rosenzweig 1995) and excluded apparent outliers according to visual assessment. The 0.06 km^2-plots were all mapped by A. Wania (Wania et al. 2006) and should not differ in mapping intensity. Therefore, no species-area curves were used for the plots.

According to the large-scale analyses in Chapter I, trait data originate from BiolFlor (Klotz et al. 2002; Kühn et al. 2004b; http://www.ufz.de/biolflor) and LEDA (Kleyer et al. 2008; http://www.leda-traitbase.org/LEDAportal).

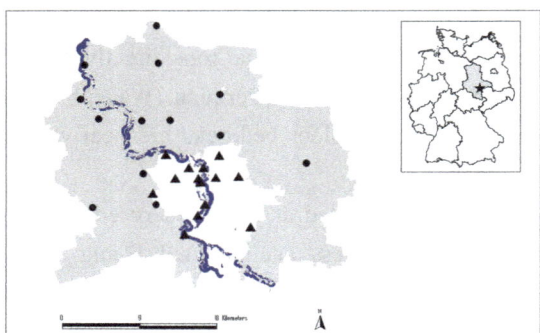

Figure 2. 1 – Protected area > 0.06 km² in the city of Halle and the adjacent rural former district of Saalkreis in Central Germany

Triangles indicate protected areas in the city of Halle (which is shown in white), dots indicate protected areas in the rural district of the former Saalkreis (which is shown in grey). The river Saale is shown in blue. The smaller picture shows the location of Halle (black star) in Central Germany (federal state of Saxony-Anhalt, shown in grey).

Because the small scale is climatically more or less uniform, we did not use climatic variables besides the land-use categories urban and rural to explain the functional composition of the flora like we did in Chapter I. Instead, we used edaphic parameters and data on habitat and land-use types to correct for other effects than differences between urban and rural conditions (for an overview see Table 2.1; Environmental Agency Saxony-Anhalt 1997; Stadt Halle 2003b; data for the 0.06 km²-plots originate from field sampling and official habitat maps and were provided by A. Wania). For the protected areas, we chose parameters whose importance for vascular plant species richness was shown by Knapp *et al.* (2008a). For the 0.06 km²-plots, we chose parameters whose importance for plant species richness was shown by Wania *et al.* (2006).

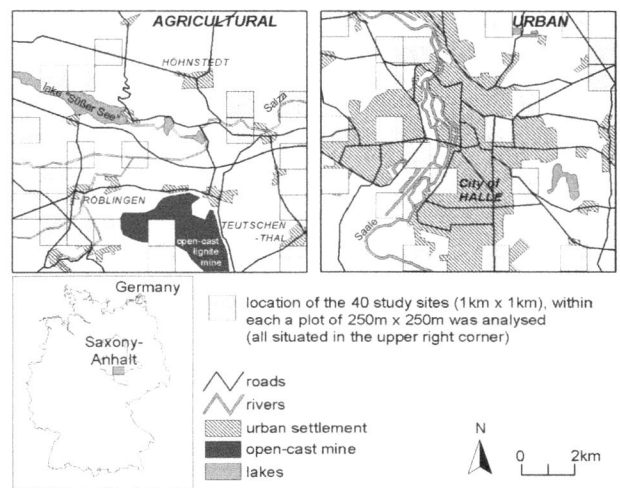

Figure 2. 2 – 40 randomly selected study sites in the city of Halle and the adjacent rural former districts of Saalkreis and Mansfelder Land in Central Germany
20 study sites each are located in Halle and its adjacent districts, respectively. The smaller picture shows the location of Halle, Saalkreis and Mansfelder Land in Germany. Figure taken from Wania *et al.* (2006).

2.3. Data Analyses

By combining the matrices on species per study site and trait state per species, we calculated the proportion of trait states per study site. In accordance with Chapter I, we calculated the log-transformed ratio of one trait state to another (e.g. Aitchison 1982; Elston *et al.* 1996; Kühn *et al.* 2006) and explained the log-ratios in linear models. The position of a study site in the city of Halle or its rural surroundings was the categorical predictor; the parameters on soils and habitat- and land-use types (Table 2.1) were continuous predictors. In the model output of linear models, the first level of a categorical predictor appears as the intercept of the linear model. For the other level, the parameter estimate shows whether it differs from the first level (here, levels are 'urban' and 'rural'). Each model was reduced via backward selection until achieving its minimal adequate version. The different models were compared by AIC (Mac Nally 2000).

We excluded one urban protected area from the analysis due to its low mapping intensity. The number of protected areas thus reduced to 13 urban and rural ones, respectively. We performed all these analyses with the open source software R, Version 2.6.0 (R Development Core Team 2007).

Table 2. 1 – Parameters on soils and habitat and land-use types used to explain differences in the functional composition of the flora in protected areas and randomly selected 0.06 km²-plots in Halle, Saalkreis and Mansfelder Land in Central Germany

Parameter	Protected areas	0.06 km²-plots
Soils	Number of patches[1]	
Habitat- and land-use types with the categories	Mean perimeter-to-area ratio[2] Number of patches[2]	Coefficient of patch-size variation[3]
Agriculture, gardens, vineyards		Contrast-weighted edge density[3]
Built-up area		Edge-density[3]
Public parks		Mean patch size[3]
Vegetation-free area		Number of edges[3]
Water bodies		Number of patches[3]
Herbaceous vegetation		Number of types[3]
Grove		
Forest		
Undefined		

1 Environmental Agency Saxony-Anhalt 1997; 2 Stadt Halle 2003a; 3 Wania *et al.* 2006

3. Results

There were only few differences between urban and rural protected areas or urban and rural 0.06 km²-plots. Wind-dispersed species had decreased proportions in urban study sites at the favor of dysochorous, endozoochorous, and hemerochorous species. The proportion of species with hygromorphic leaves was increased in urban 0.06 km²-plots at the expense of plants with scleromorphic leaves. Chamaephytes were less frequent in urban than rural protected areas in comparison to therophytes. Pluriennials were more frequent in urban than rural 0.06 km²-plots in comparison to biennials. Specific leaf area was higher in urban than rural study sites (both protected areas and 0.06 km²-plots). Plants in urban protected areas less often had UV-reflecting flowers but plants in urban 0.06 km²-plots had more often UV-reflecting flowers than plants in rural sites, respectively (for all results see Table 2.2).

4. Discussion

Why are there relatively few differences between urban and rural study sites in the city of Halle and its rural surroundings but many differences between urbanized and rural grid-cells in Germany on a large scale (Chapter I)? First, this migh be a statistical artifact: Sample size on the large scale is larger than sample size on the small scale (cf. Table A2). However, the power of the z-statistic used for the comparison of urbanized and non-urbanized grid-cells is independent of sample size. Moreover, the same comparison performed with linear models (Knapp *et al.* 2008b) yielded the same results like the z-statistic. Still, we used linear models for the small-scale analyses and sample size on the small scale might be too small to yield meaningful results. Nevertheless, it is more likely that differences in the occurrence of rare species in large- and small-scale study sites are responsible for the lack of many large-scale patterns on the small scale: The chance that a rare species that is present in the 130 km²-grid-cell including Halle is also present in any one of the protected areas or 0.06 km²-plots is small. Remember that there are generally more rare than common species (Gaston & Blackburn 2000) and that every species has the same influence on the trait state patterns, because we only used presence-absence data but no abundance data. Rare native species might be present in the protected areas, but many neophytes are also rare (Williamson & Fitter 1996; Hulme 2008) and probably restricted to typical urban-industrial habitats. Thus, especially plants with pre-adaptations to urban land-use, which should be the ones that cause the large-scale patterns, might not occur in Halle's protected areas.

Secondly, Halle and its surroundings are situated within one biogeographic region, in contrast to the grid-cells analyzed in Chapter I. However, land use explained even more variation than biogeography at the scale of grid-cells; thus, the urban-rural gradient should also cause differences in functional composition on a small scale. Instead of biogeography, effects of nature conservation might hide trait state patterns: Halle's protected areas are remnants of semi-natural landscape, e.g. dry lawns on porphyric rocks or alluvial forests, which are more similar to the protected areas in the rural district than urbanized, agricultural and semi-natural grid-cells in Germany are to each other.

In summary, the lack of trait state patterns in the study sites in Halle and surroundings mainly seems to be due to effects of nature conservation and the stochastic lack of rare species in the relatively small study sites.

The type of dispersal, which shows similar differences between urban and rural study sites like between urbanized and rural grid-cells, is related to fragmentation: Plants in isolated habitat patches depend on vectors that spread their seeds to reach other habitat patches. Urbanized areas are especially fragmented, consisting of many different patches, with similar patches like woodland-patches or meadow-patches scattered over the city (Niemelä 1999). Consequently, traits that help to cope with fragmentation are of special importance for the persistence of plants in urbanized areas. Our results suggest that wind-dispersal is less suitable in urbanized areas than animal- or human-dispersal, like the large-scale analysis already revealed. On the one hand, this might be due to calms (Kuttler 1993), wind channeled in streets and habitats in the lee of buildings that wind cannot reach; on the other hand, the high human population density should be positive for both animal- and human-dispersed species (von der Lippe & Kowarik 2007; 2008; see *Discussion* in Chapter I).

Plants with hygromorphic and scleromorphic leaves show opposite patterns on the large and the small scale with relatively more hygromorphic plants in the study sites in Halle than in its rural surroundings: Gardens, parks and old cemeteries in the city of Halle are often shady and should favor plants with hygromorphic leaves, while the 0.06 km²-plots in the rural districts are mostly situated outside of villages and thus often in treeless, unshaded agricultural areas. More-over, several of the urban 0.06 km²-plots are located near the river Saale, where air moisture is increased in spite of the surrounding urban matrix and where alluvial forests provide shady habitats. Some other urban 0.06 km²-plots are located in the 'Dölauer Heide' (see Fig. 4.1, Chapter IV), a relatively large for-ested (and partly protected) area in the northwestern part of the city, where hy-gromorphic plants also should be favored by the cooling shadow of trees. These effects seem to be strong enough to locally decrease the influence of low air moisture in open or built-up urban habitats and cause the pattern opposing the large-scale results, where plants with hygromorphic leaves were less frequent in urbanized than in non-urban grid-cells.

Also biennial and pluriennial plants show opposing patterns on the large and the small scale, with relatively more pluriennials in urban 0.06 km²-plots. This might as well be due to the 'Dölauer Heide' and the alluvial forests because forested areas are rare in the rural surroundings of Halle, making the 'Dölauer Heide' within the city one of the largest forested areas in the region. Disturbance intensity surely is reduced in the forest in comparison to the urban matrix in general, fostering pluriennial species. Also the meadows in the urban parks and gardens are habitats of pluriennial species. At the other hand, agriculture is the dominating land use of Halle's environs, fostering short-lived species (cf. Losos-ová *et al.* 2006). However, phanerophytes, although all being pluriennial, show no differences between urban and rural study sites, indicating that a similar num-ber of tree and shrub species grow in both urban and rural study sites. The inclu-sion of abundance data might clarify this pattern.

Chamaephytes and therophytes in the protected areas resemble the large-scale patterns with increased proportions of therophytes. Vegetation in some of the protected areas in Halle probably gets disturbed more often than in the rural protected areas because urban green spaces provide the only recreation area nearby for many urban dwellers. Several of the urban protected areas are located in the Saale valley near the city center and are very well frequented. Walkers, people doing sports or walking their dogs are likely sources of disturbance.

Table 2. 2 – Differences in trait state ratio between urban and rural protected areas and urban and rural randomly selected 0.06 km²-plots in Halle, Saalkreis, and Mansfelder Land in Central Germany. See below for further explanation.

Trait	Trait state ratio	Protected areas			0.06 km²-plots		
		Model R^2	Int_u	Int_{rur}	Model R^2	Int_u	Int_{rur}
Dispersal type	Dysochorous/Anemochorous	0.24*	-0.14	-0.32*	0.15*	-0.24	-0.42 n.s.
	Endozoochorous/Anemochorous	0.16*	0.28	0.18*	0.05 +	n.s.	n.s.
	Epizoochorous/Anemochorous	0.08 n.s.	0.47	0.55 n.s.	0.1*	n.s.	n.s.
	Hemerochorous/Anemochorous	0.23**	0.53	0.34**	0.14*	0.7	0.53 +
	Hydrochorous/Anemochorous	NULL***	n.s.	n.s.	0.3***	n.s.	n.s.
Leaf anatomy	Helomorphic/Hygromorphic	NULL***	n.s.	n.s.	0.07 n.s.	n.s.	n.s.
	Hydromorphic/Hygromorphic	NULL***	n.s.	n.s.	NA	NA	NA
	Mesomorphic/Hygromorphic	0.09 +	n.s.	n.s.	0.09 +	1.72	2.0 n.s.
	Scleromorphic/Hygromorphic	0.12*	n.s.	n.s.	0.18**	0.32	0.95**
	Succulent/Hygromorphic	0.04 n.s.	n.s.	n.s.	0.52***	n.s.	n.s.
LDMC	-	0.02 n.s.	n.s.	n.s.	0.17**	n.s.	n.s.
Leaf persistence	Evergreen/overwintering green	0.11*	n.s.	n.s.	0.33***	n.s.	n.s.
	Spring green/overwintering green	NULL***	n.s.	n.s.	NA	NA	NA
	Summer green/overwintering green	0.11 +	n.s.	n.s.	0.18*	n.s.	n.s.
Life form	Chamaephyte/Therophyte	0.33**	-1.49	-1.24 +	0.21**	n.s.	n.s.
	Geophyte/Therophyte	0.13*	n.s.	n.s.	0.23*	n.s.	n.s.
	Hemicryptophyte/Therophyte	0.28**	n.s.	n.s.	0.24***	n.s.	n.s.
	Hydrophyte/Therophyte	NULL***	n.s.	n.s.	0.62***	n.s.	n.s.
	Phanerophyte/Therophyte	0.18*	0.41	0.13 n.s.	0.46***	n.s.	n.s.
Life span	Annual/pluriennial	0.27**	n.s.	n.s.	0.42***	n.s.	n.s.
	Biennial/pluriennial	NULL***	n.s.	n.s.	0.41***	-1.21	-0.94*
Pollen vector	Insects/wind	0.05 n.s.	n.s.	n.s.	0.21*	n.s.	n.s.
	Selfing/wind	NULL***	n.s.	n.s.	0.12*	n.s.	n.s.
SLA	-	0.17*	27.57	25.75 +	0.33***	29.64	27.73*
Reproduction	Generatively only/vegetatively only	NULL***	n.s.	n.s.	0.42***	n.s.	n.s.
	Generatively & vegetatively/vegetatively only	NULL***	n.s.	n.s.	0.48***	n.s.	n.s.
UV-reflection	No/yes	0.11 +	0.71	0.61 +	0.32***	0.52	0.66 +

Further explanation for Table 2.2.: Shown are the intercepts (Int_u for urban sites, Int_{rur} for rural sites) of the linear models. For Int_{rur}, P-values show whether the trait state composition of rural study sites differs from urban study sites (i.e. from Int_u). $P \geq 0.1$ n.s., $0.1 > P > 0.05+$; $P \leq 0.05*$; $P \leq 0.01**$, $P \leq 0.001***$. Where no intercepts are given but only n.s., land use (urban/rural) was not included in the minimal adequate model (see in the *Data analyses* section of this Chapter). NULL indicates a null model without predictors. NA indicates ratios that were not calculated due to insufficient number of species (see Table A2). Model R^2 (adjusted for number of predictors) gives the total explained variance of the model including P-values.

UV-reflection shows opposing patterns in protected areas and 0.06 km²-plots, with less UV-reflecting flowers in the urban than rural protected areas (opposing the large-scale pattern), and more UV-reflecting flowers in urban than rural 0.06 km²-plots (reflecting the large-scale pattern). As the UV-reflection of flowers is mainly documented for insect-pollinated plants in the BiolFlor database (see Chapter I), it seems that insect-pollinated plants have to struggle harder to attract their pollinators in urban than rural 0.06 km²-plots, respectively, as already proposed in Chapter I. Probably, pollinating insects find better living conditions and more appropriate habitats in the protected areas, which should partly buffer influences from the surrounding urban matrix.

Summing up, although both protected areas and 0.06 km²-plots are influenced by an urban or rural environment respectively, many differences that exist between urbanized and non-urban grid-cells on the large scale are not present on the small scale (there are e.g. no urban-rural small-scale patterns for leaf persistence, pollen vector or type of reproduction). The 0.06 km²-plots were selected randomly and are scattered over the city; they partly cover semi-natural habitats like parts of the Saale valley or forested areas as well, like the protected areas (see Fig. 2.2). This suggests that semi-natural habitats within an urban matrix, e.g. protected areas but also man-made habitats like parks, cemeteries or gardens are able to support plants with trait states that are more frequent in non-urban areas on larger scales. Additionally, the lack of rare species in the relatively small study plots, be it rare neophytes, rare ruderal species of typical urban-industrial habitats or rare native species, might cause the lack of differences found on the large scale.

Of course, the comparison of protected areas and 0.06 km²-plots in Halle and surroundings is only one case study and its results might not be generally valid. However, this case study is an example of how nature conservation can contribute to a diverse functional composition of urban floras: The coexistence of typical urban habitats such as brownfields, railway sites or industrial sites and of semi-natural habitats such as protected alluvial forests or dry lawns on porphyric rocks within the same urbanized area might support a high functional diversity. The former should provide habitats for typical urban ruderal plants but also for rare plant species that lost their habitats in rural cultivated landscapes (Lenzin *et al.* 2007); the latter might even provide habitats for 'urban avoiders' (Blair 1996) with 'rural trait states'. However, area size always is a limiting factor.

Chapters I and II showed that urbanization causes shifts of species' trait state frequency on both large spatial scales and along small-scale urbanization gradients.

Chapter III – How Species Traits and Affinity to Urban Land Use Control Plant Species Frequency

1. Introduction

Urbanized areas are richer in plant species than their rural surroundings (e.g. Walters 1970; Haeupler 1975; Hope *et al.* 2003; Kühn *et al.* 2004a; Wania *et al.* 2006), due to various reasons listed in the General Introduction of this book, such as geological and structural heterogeneity, introduction of species or high temperatures. However, urbanization especially can cause the extinction of rare native species by altering or destroying natural and semi-natural habitats. The high urban species richness is based on both alien and common native species (Kühn *et al.* 2004a) and a few species rich lineages (see Chapter V), at least in Germany. We consequently need strategies for nature conservation in urban areas that sustain a high species diversity including rare and endangered species (Schwartz *et al.* 2006). We then need to know why species are rare and how urban land use influences rarity.

Within the past decades, plant species frequency (or rarity) has become one of the most important parameters used in nature conservation when assessing the threat status of species and when making management and conservation decisions (Gaston 1994; Dobson *et al.* 1995). Generally, actual species distribution has been taken as an indicator for species frequency (e.g. Ellenberg *et al.* 2001) and the determination of species rarity is mostly based on grid-based abundance measures (e.g. Dony & Denholm 1985; Kunin 1998). This approach may, however, be too simple as species' distribution ranges are correlated to their niche width and specializations (Rabinowitz 1981; Ehrlén & Eriksson 2000). A species showing low absolute frequency may be rare because it occurs in rare habitats (Rabinowitz 1981) or it may be rare for other reasons although its habitat is frequent (Römermann *et al.* 2007). Hence, it is important to consider habitat frequencies as well, i.e. the number of grid-cells where the respective habitat occurs, when using species frequencies to measure species rarity. Otherwise, the frequency of "naturally rare species" sensu Rabinowitz (1981), i.e. species occurring only in rare habitats, would be underestimated. We follow this argumentation by using relative species frequencies as a measure for species rarity (Römermann 2006).

Besides habitat availability, species rarity is influenced by the species' adaptability to environmental changes and hence by their life-history traits: Some traits make species more extinction-prone, others less (Poschlod et al. 2000; Cardillo et al. 2004). Römermann et al. (2008), for example, showed that species of dry grasslands are mainly rare when preferring warm, dry, light and nutrient-poor conditions. Other studies identified traits that are related to the rarity of insect visited forbs (Bekker & Kwak 2005) or to the regional frequency of forest herbs (Matlack 2005).

Traits alone, however, seem insufficient to fully explain species rarity; interactions between species traits and environmental factors related to land-use change, e.g. elevated temperature or disturbance regimes, might yield more meaningful results (cf. Fréville et al. 2007).

So far we are aware of only two studies that related species extinction risk to both urban land use and life-history traits: Preston (2000) compared historical species extinctions in two British counties, one dominated by agriculture, the other by urban land use. He found that both land-use types preferably threaten small species of open, unfertile habitats. Williams et al. (2005) showed for Western Victoria in Australia that urbanization increases the extinction risk of grassland species that were geophytes or hemicryptophytes with a flat rosette and dispersed by wind or ants.

We will go one step further by investigating whether species frequency is related to functional traits while accounting for the affinity of plants to urban land use (called 'urbanity' hereafter) without focusing on one specific habitat type. Specifically, we will explain the relative frequency of vascular plant species in Germany with traits relevant for dispersal, persistence and reproduction and we will include the interactions of these traits with urbanity to assess whether a trait makes a plant more or less frequent when having high urbanity. We will also account for the species' phylogenetic relationships. On the one hand, native species richness in German cities is based on common species (Kühn & Klotz 2006), thus there should be a positive relation between species affinity to urban land use and species frequency. On the other hand, many alien species have been introduced into urbanized areas (Kent *et al.* 1999), most of them being rare in their area of introduction (Hulme 2008), at least at the beginning of their invasion process. Thus, species affinity to urban land use might also be negatively related to species frequency. However, as there are far more native than alien species in the German flora, we suppose that species with a high affinity to urban land use are relatively frequent, while species with a low affinity to urban land use are less frequent, reflecting the loss of rare species and the gain in common species in urbanized landscapes.

2. Materials and Methods

2.1. Data Sources

2.1.1. Species Rarity

In accordance with Chapter I, plant species occurrences per 12 km × 11 km grid-cell were taken from FLORKART, the database of the German flora. Again, we only used grid-cells with at least 45 of 50 control species (Kühn et al. 2004a; 2006). Furthermore, we only used grid-cells with at least 50% area in Germany, resulting in 2709 grid-cells for analyses.

We used all species (including the control species) except those without phylogenetic classification and aquatic species (the latter as defined by Korneck et al. 1998 according to their main habitat). Aquatic habitats possess other conditions than terrestrial habitats, thus other traits should be relevant for the frequency of aquatic species. Furthermore, we excluded all *Rubus* species but *R. caesius* L., *R. chamaemorus* L., *R. armeniacus* Focke, *R. laciniatus* Willd., *R. idaeus* L., *R. phoenicolasius* Maxim., *R. saxatilis* L. and *R. spectabilis* Pursh. All other 148 *Rubus* species in our original species list but the two are apomictic microspecies, i.e. they are very similar to each other and can be seen as pseudoreplicates. In total, 1776 species were included in the analyses.

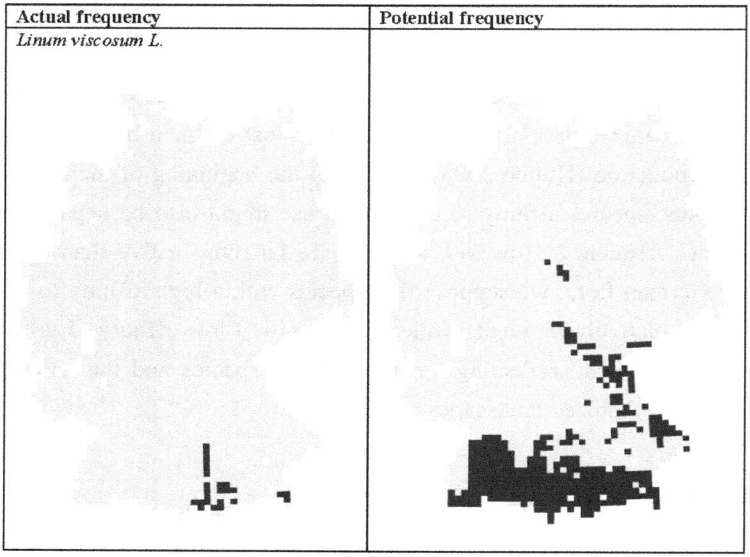

Actual frequency	Potential frequency
Linum viscosum L.	

Figure 3. 1 – Comparison of actual and potential frequeny
The example shows *Linum viscosum L.* Actual frequency is the actual occurrence of the species per grid-cell (from FLORKART); potential frequency was modeled with parameters on climate, topography, soils, geology, and the occurrence probability of habitat-specific species in BIOMOD (Thuiller 2003; see *Materials and Methods*). The number of cells for actual frequency divided by the number of cells for potential frequency yields the relative frequency of the plant species.

We used relative species frequency to measure species rarity, i.e., the grid-based actual occurrence of a species divided by its potential grid-based occurrence (Fig. 3.1; see Römermann 2006 for a detailed description of actual and relative frequencies). To obtain the relative frequency of a plant species, we first had to estimate the potential range. Therefore, we modeled the potential occurrence of each species with several parameters on land use, climate, topography, soils, geology and the co-occurrence of habitat-specific species (Table 3.1) in an ecological niche modeling approach (Guisan & Thuiller 2005) for two thirds of all grid-cells, using generalized linear models (GLM).

Data on land use were again taken from Corine land cover data (Statistisches Bundesamt 1997). We calculated the proportions of agricultural, forested/semi-natural land-use classes and wetlands per grid-cell, but omitted the land-use class water bodies according to the omission of aquatic species. We also omitted the land-use class of artificial land-use types because we used them later to calculate the affinity of species to urban land use. This exclusion avoided circular reasoning.

Data on climate, topography, soils and geology originate from various sources and were available per grid-cell (see Table 3.1 and references therein). At the time of the analyses for Chapter III, improved climatic parameters were available per grid-cell from Badeck *et al.* (2008), which had not yet been available at the time of analyses for Chapters I and V. Therefore, the climatic parameters differ from the parameters used in Chapters I and V. In addition to these parameters, we included some proxies for habitat conditions by calculating the probability of occurrence of habitat-specific species (Korneck *et al.* 1998) following the approach described in Römermann *et al.* (2007). In short, it represents the probability that a group of species characteristic for one specific habitat occurs in a grid-cell as estimated from specific species co-occurring in the grid-cells. However, we did not calculate occurrence probability for all 24 groups of habitat-specific species which are distinguished by Korneck *et al.* (1998); instead, we divided them into three groups representing (i) habitats defined climatically or edaphically (e.g. dry lawns or bogs), (ii) habitats defined geographically (e.g. from coasts or high mountains), and (iii) common habitats (e.g. arable fields or deciduous forests). These three groups were derived from the comparison of two alternatives of our modeling approach (not shown; Römermann *et al.* unpublished): One used land use, climate, topography, soils and geology to explain species actual occurrences, the other used the co-occurrence of habitat-specific species from all the 24 groups defined by Korneck *et al.* (1998) as explanatory variables. This comparison yielded three distinct groups of species; one estimated similar by both approaches (species of habitats defined climatically or edaphically), one underestimated by the first approach (species of habitats defined geographically), and the third underestimated by the latter approach (species of common habitats). Based on this comparison, we used the parameters on land use, climate, topography, soils and geology together with the occurrence probability of the three groups of species as explanatory variables to reduce the risk of under- or overestimation of species occurrences.

The models that explained the actual occurrence of species for two thirds of grid-cells were now validated by applying them to the same species but for the remaining third of grid-cells. The agreement between the actual occurrence of a species and the potential occurrence of this species predicted by the models was evaluated by AUC (area under the curve, Thuiller *et al.* 2003), which is not dependent on a specific probability threshold. We applied the models to all grid-cells to calculate the potential frequency of each species for the whole study area. We performed these calculations in BIOMOD (Thuiller 2003).

2.1.2. Species Affinity to Urban Land Use

To investigate how species traits and affinity to urban land use (urbanity) jointly control relative species frequency, we calculated urbanity as Pearson's *r* for the correlation of species' actual occurrence per grid-cell and the intensity of urban land use per grid-cell, according to the percentage of Corine artificial land-use types (for a similar approach see Thompson & McCarthy 2008). Corine artificial land-use types include built-up residential, industrial, commercial and transport area, mines, dumps, and artificial non-agricultural vegetated areas, i.e. urban green.

2.1.3. Species Traits and Phylogeny

Species traits were taken from BiolFlor (Klotz *et al.* 2002; Kühn *et al.* 2004b), LEDA (Kleyer *et al.* 2008), and the Seed Information Database (Flynn *et al.* 2004). We chose traits related to plant performance in situ (e.g. leaf traits), reproduction (e.g. pollination type), dispersal (e.g. dispersal type), persistence (e.g. seed mass), and plant origin (e.g. floristic status; see Table A1). Categorical traits were translated into dummy variables prior to analyses, because plants can have several states of one trait, such as plants that are both insect- and self-pollinated.

Hence, we changed the original matrix containing 17 traits in a matrix with seven numerical traits and 43 trait states deduced from ten nominal traits.

We also included the Ellenberg values for moisture, nitrogen and temperature (Ellenberg *et al.* 2001). Ellenberg values are species-specific scores ranging from 1-9 (or 1-12 for moisture). They estimate the optimum ecological occurrence of species along environmental gradients, considering competition between species and reflecting habitat conditions, i.e. they reflect the realized niche of a species (Ellenberg *et al.* 2001). Ellenberg values behave as continuous variables for sample sizes exceeding 100 species and can be used as numerical variables (Ter Braak & Barendregt 1986). However, according to the omission of aquatic species, we excluded the trait states "hydrophyte" and "hydromorphic leaves".

The phylogenetic relationships of species are documented in BiolFlor. From the phylogenetic code, we determined the number of nodes separating one species from another but not the length of the branches of the respective phylogenetic tree. However, the number of nodes approximates the lengths of the branches which are set to unity (Faith 1992).

2.2. Data Analyses

We explained relative species frequency with the interactions between traits and urbanity first in GLMs, then in simultaneous autoregressive error models (SARerr; see Dormann *et al.* 2007; Kissling & Carl 2008) to correct for phylogenetic relationships of species. In the SARerr-models, we included the topology of our phylogenetic tree instead of spatial information, defined species with up to 15 nodes distance to each other as neighbors, thus turning a spatial model into a phylogenetic model.

Table 3. 1 – Environmental parameters used to calculate the potential and relative frequency of vascular plant species in Germany

Reference period for all climatic parameters: 1961-1990

Environmental parameter	References
Mean temperature of the warmest month	Badeck *et al.* 2008
Mean temperature of the coldest month	
Temperature range per year	
Number of frost days per year	
Precipitation variability	Provided by the German Meteorological Service (Deutscher Wetterdienst, Department Klima und Umwelt; cf. Kühn *et al.* 2003)
Potential evapotranspiration	Badeck *et al.* 2008
Mean height above sea level	ARCDeutschland500 dataset, scale 1:500,000, provided by ESRI; (cf. Kühn *et al.* 2003)
Mean wind speed	Provided by the German Meteorological Service (Deutscher Wetterdienst, Department Klima und Umwelt; cf. Kühn *et al.* 2006)
Number of soil patches	German soil survey map 1:1,000,000 (Bundesanstalt für Geowissenschaften und Rohstoffe 1995)
Number of soil types	
Number of geological patches	German geological survey map 1:1,000,000 (Bundesanstalt für Geowissenschaften und Rohstoffe 1993)
Number of geological types	
Proportion of agricultural land-use types	Corine land cover data (Statistisches Bundesamt 1997)
Proportion of forested / semi-natural land-use types	
Proportion of wetlands	
Probability of occurrence of edaphically/climatically determined habitats[1]	Römermann *et al.* 2007 using data from Korneck et al., 199); http://www.floraweb.de/
Probability of occurrence of geographically determined habitats[1]	
Probability of occurrence of common habitats[1]	

1 see *Materials and Methods* section of this Chapter for a more detailed description

We first performed this procedure for each of the seven numerical traits and 43 trait states separately in order to identify trait-urbanity interactions and tested the models' residuals for normal distribution with Kolmogorov-Smirnov tests.

We then included all significant single trait-urbanity interactions ($p < 0.05$ in SARerr-models) into multi-trait GLMs. Additionally, we included the trait states "neophyte" and "therophyte" that had no significant urbanity-interactions but for which we expected interactions in the multi-trait GLMs: Both occur preferentially in anthropogenic habitats; especially neophytes are mainly dispersed by humans and are over-represented in cities (Kühn et al. 2004a). We further included the Ellenberg values for moisture, nitrogen and temperature. They clearly differ between urban and rural areas (Wittig & Durwen 1982) and can add valuable information about species rarity (Römermann et al. 2008).

To assess the importance of each trait within the multi-trait model we developed several full models for the 700 species with full trait documentation; one model containing all traits chosen for multi-trait GLMs, the others containing all but one of the traits, with each trait excluded in one of the models, plus one model excluding urbanity, containing the traits without interactions, resulting in 25 different full models explaining relative species frequency. With this procedure, we excluded random effects caused by collinearities between traits. We reduced each full model via backward selection of least significant variables, using the Akaike Information Criterion (AIC) for model comparison (Mac Nally 2000). The minimal adequate models (called 'multi-trait models' hereafter) were then included into SARerr-models to check whether correcting for phylogenetic relationships changed results. Variables occurring in at least ten of the multi-trait models, i.e. the variables most affecting relative species frequency, were included in a final model, which was again corrected for phylogenetic relationships.

We performed all calculations with the open source software R, version 2.6.0 (R Development Core Team 2007). SARerr-models were computed in package spdep (Bivand et al. 2007).

3. Results

3.1. Single-Trait Models

In the single-trait models to explain relative species frequencies, seed mass, canopy height (both log-transformed), hemerochory, regular and scarce leaf distribution, summer green leaves, phanerophytes, wind-pollination, and global distribution had significantly negative urbanity-interactions (Table 3.2), i.e. the higher the affinity of a plant species with one of these traits to urban land use is, the lower is its relative frequency. Anemochory, hemirosettes, evergreen leaves, hemicryptophytes, insect- and self-pollination, archaeophytes, temperate-meridional distribution, and reproduction by seed had significantly positive urbanity-interactions. Generally, phylogenetic autocorrelation was small (mean + SE of λ = 0.0028 + 0.0002) and hardly changed the results of the GLMs.

3.2. Multi-Trait Models

In the multi-trait models, the interaction of self-pollination with urbanity influenced relative species frequency positively, showing that plants with a high urbanity are more frequent when they are capable of self-pollination (Fig. 3.2a; Table 3.3). Similarly, the interaction of reproduction by seed with urbanity had a positive effect on relative species frequency, i.e. plants that reproduce by seeds are more frequent the higher their urbanity (Fig. 3.2b; Table 3.3). Moreover, the interaction of Ellenberg temperature with urbanity had a negative effect on relative species frequency, showing that species with medium temperature preferences are more frequent than species with high temperature preferences (Fig. 3.2c; Table 3.3). Ellenberg moisture interacted negatively with urbanity in four of the models (the ones excluding hemerochory, temperate-meridional distribution, self-pollination or reproduction by seed; Fig. 3.2d; Table 3.3). Hemirosettes interacted positively with urbanity in five of the multi-trait models (the ones excluding canopy height, Ellenberg- moisture or Ellenberg-temperature, neophytes of self-pollination; Fig. 3.2e; Table 3.3). Evergreen leaves interacted with urbanity in only one model, being positively related to urbanity when Ellenberg temperature was excluded (Table 3.3).

Table 3. 2 – Results of the single-trait general linear models and the spatial autoregressive error models carried out on relative species frequencies in Germany and the relevance of life-history traits and plant affinity to urban land use. See below for further explanation.

Traits / Trait states	N	GLM Model	GLM Trait	GLM URB	GLM Trait:URB	SARerr λ	SARerr Trait	SARerr URB	SARerr Trait:URB
Log(Canopy height)	1604	***	*** (+)	n.s.	*** (-)	0.004***	***	n.s.	***
Anemochory	1488	**	*** (+)	n.s.	**	-0.001 n.s.			
Chamaechory	1488	n.s.	n.s.	n.s.	n.s.	0.001 n.s.			
Dysochory	1488	***	*** (+)	n.s.	n.s.	0.002 n.s.			
Endozoochory	1488	***	*** (+)	+	n.s.	0.003 n.s.			
Epizoochory	1488	**	*** (+)	n.s.	n.s.	0.000 n.s.			
Hemerochory	1488	***	*** (+)	n.s.	* (-)	0.002 n.s.			
Hydrochory	1488	***	*** (+)	n.s.	n.s.	0.002 n.s.			
Ellenberg_F	1516	***	*** (+)	n.s.	n.s.	0.001 n.s.	***		
Ellenberg_N	1476	***	*** (+)	n.s.	n.s.	0.005***	***	n.s.	n.s.
Ellenberg_T	1341	***	*** (-)	n.s.	n.s.	0.0002 n.s.			
Native	1712	***	*** (+)	** (+)	n.s.	0.003+	***	**	n.s.
Archaeophyte	1712	***	n.s.	n.s.	*** (+)	0.003*	n.s.	n.s.	**
Neophyte	1712	***	*** (-)	*** (+)	n.s.	0.003*	***	***	n.s.
FZ_allrounder	1689	***	*** (+)	* (+)	* (-)	0.003*	***	**	*
FZ_extratropic	1689	***	*** (+)	*** (+)	n.s.	0.003*	***	***	n.s.
FZ_meridional	1689	***	*** (-)	*** (+)	n.s.	0.003+	***	***	n.s.
FZ_temperate	1689	***	** (-)	* (+)	n.s.	0.003*	**	*	n.s.
FZ_temperate-north	1689	***	*** (-)	* (+)	n.s.	0.004*	***	*	n.s.
FZ_temperate-meridional	1689	***	*** (-)	* (-)	*** (+)	0.003*	***	+	***
Helomorphic leaves	1598	*	** (+)	n.s.	n.s.	0.003*	*	n.s.	n.s.
Hygromorphic leaves	1598	***	*** (+)	n.s.	n.s.	0.004**	***	n.s.	n.s.
Mesomorphic leaves	1598	n.s.	n.s.	n.s.	n.s.	0.004*	n.s.	n.s.	n.s.
Scleromorphic leaves	1598	*	** (-)	n.s.	n.s.	0.004**	**	n.s.	n.s.
Succulent leaves	1598	*	** (-)	n.s.	n.s.	0.003+	*	n.s.	n.s.

Table 3.2. - continued

Traits / Trait states	N	GLM				SARerr			
		Model	Trait	URB	Trait:URB	λ	Trait	URB	Trait:URB
LD_regular	1750	***	***(+)	**(+)	**(-)	0.004**	***	**	*
Rosettes	1750	n.s.	n.s.	+(+)	n.s.	0.003*	n.s.	*	n.s.
Hemirosettes	1750	***	***(-)	n.s.	**(+)	0.004*	***	n.s.	*
LD_scarce	1750	***	+(+)	*(-)	*(-)	0.003+	+	*	*
Evergreen leaves	1632	***	*(+)	n.s	+(+)	0.004**	*	n.s.	*
Springgreen leaves	1632	n.s.	n.s.	n.s	n.s.	0.004*	n.s.	*	n.s.
Summergreen leaves	1632	***	+(-)	+(+)	*(-)	0.004**	+	***	**
Wintergreen leaves	1632	+	n.s.	n.s	n.s.	0.004**	n.s.	n.s.	n.s.
Chamaephytes	1746	n.s.	n.s.	n.s	n.s.	0.003*	n.s.	+	n.s.
Geophytes	1746	n.s.	n.s.	+(+)	n.s.	0.003*	n.s.	*	n.s.
Hemicryptophytes	1746	***	**(+)	n.s	*(+)	0.002+	**	n.s.	*
Phanerophytes	1746	+	n.s.	*(+)	+(-)	0.003*	n.s.	**	+
Therophytes	1746	+	n.s.	*(+)	n.s.	0.003+	n.s.	*	n.s.
Short-lived	1746	n.s.	n.s.	n.s.	n.s.	0.002+	n.s.	n.s.	n.s.
Perennial non-clonal	1746	n.s.	n.s.	n.s.	NA	0.002+	n.s.	n.s.	NA
Perenn-clonal	1746	n.s.	n.s.	n.s.	n.s.	0.002			
LDD	1488	***	***	n.s.	n.s.	0.003			
Insect-pollination	1680	*	n.s.	n.s.	*(+)	0.003+	n.s.	n.s.	*
Self-pollination	1680	*	n.s.	n.s.	+(+)	0.004*	n.s.	n.s.	+
Wind-pollination	1680	**	**(+)	**(+)	*(-)	0.002	n.s.	n.s.	
Log(Seed mass)	1305	**	n.s.	n.s.	**(-)	0.004+	n.s.	n.s.	***
SLA	1287	***	***(+)	n.s.	n.s.	0.003	n.s.		
Seed reproduction	1746	*	n.s.	n.s.	*(+)	0.003*	n.s.	n.s.	*

Further explanation for table 3.2.: Results of the Kolmogorov-Smirnov-test for all model outputs: $P > 0.9$. Number of species included in the models (N), the results of the F-statistics of the general linear model (GLM, Model). Significance of parameter estimates of the respective trait ("Trait"), urbanity (URB) and their interaction (Trait:URB) are given for both the GLM that did not account for phylogeny and for the spatial autoregressive error model (SARerr) that accounted for phylogeny. (+) and (-) in the columns "Trait", "URB" and "Trait:URB" indicate whether the significant traits, URB or the interaction were positive (+) or negative (-). λ and its significance indicate the presence of phylogenetic autocorrelation. When species data was autocorrelated, parameter estimates of the SARerr model were given. GLM: urbanity vs. relative species frequency: $R^2=$ 0.002*; $\lambda = 0.003$ n.s.; NA = not available. Levels of significance: *** $P \leq 0.001$, ** $P \leq$ 0.01, * $P \leq 0.05$, + $P \leq 0.1$, n.s. not significant.

Urbanity itself was positively related to relative species frequency in every multi-trait model, except for the model excluding Ellenberg temperature (where the effect of urbanity was negative).

The traits occurring in nearly every multi-trait models (i.e. in 23 or 24 of the 25 models, see Table 3.3) and the significant interactions of urbanity with self-pollination, type of reproduction and Ellenberg temperature were then included in the final model (Table 3.4). This model could not be reduced, although still containing non-significant variables, because deleting these variables increased the AIC and thus decreased the model fit (see Table 3.4 for the parameter estimates within the final model). Phylogenetic autocorrelation did not change results (not shown).

4. Discussion

The study clearly shows that species frequency is influenced by both species traits and urbanization. The higher the affinity of plants to urban land use, the higher is their relative frequency. Besides, urbanization interacts with the species' type of pollination, type of reproduction and with the species' general habitat preferences for temperature. To a minor extent, it interacts with hemirosettes and habitat preferences for moisture.

4.1. Traits and Relative Frequencies

The following trait states influence relative species frequency positively, independent of the species' degree of urbanity (Table 3.3): canopy height, dispersal by wind or humans, having a global distribution, being a hemicryptophyte, and preferring nitrogen-rich habitats. Canopy height is related to a plant's competetiveness (Cornelissen *et al.* 2003): Higher plants shade smaller plants and might outcompete them. Species dispersed by wind or humans are capable of long-distance dispersal and less restricted by fragmentation. Moreover, human-dispersed species profit from human activities, which are not restricted to urban areas. If a species is globally distributed, it is likely able to occur in many different habitats, not only worldwide but also in Germany. Hemicryptophytes are generally strong competitors that gain in dominance in the course of succession (Raunkiaer 1934, cited in Ecke & Rydin 2000) or with increasing nutrient supply (Smart *et al.* 2005; Römermann *et al.* 2008). Plant species preferring nitrogen-rich habitats profit from the high loads of nitrogen from agriculture, traffic, and industries in European landscapes (Franzaring & Fangmeier 2006).

There are also trait states that influence relative species frequency negatively and independent of urbanity, namely being a neophyte and having a temperate-meridional distribution. It is surprising that neophytes do not interact with urbanity, although cities are the places where alien plants are transported to (via trade and traffic) and where aliens from warmer climates find temperatures high enough to persist (Sukopp *et al.* 1979). Thus, neophytes with a low urbanity should be less frequent than neophytes with a high urbanity. However, the generally low relative frequency of neophytes confirms that most neophytes are not invasive but rather rare, at least at the beginning of their invasion process (Williamson & Fitter 1996). This might change, because "newcomers" need some time to spread in a new area, being rare now but getting more common (Cadotte *et al.* 2006). Species with temperate-meridional distribution should also be rare in Germany because they are at the northern border of their range and only find suboptimal growing conditions (Korneck *et al.* 1998). However, the positive interaction with urbanity in the single-trait model (Table 3.2) suggests that 'southern' plants are already frequent in urbanized areas, while still rare in rural areas, due to the urban 'heat island' (Oke 1982) and drier conditions in urban environments. Neophytes and species with temperate-meridional distribution might become more frequent with climate warming (cf. Bañuelos *et al.* 2004).

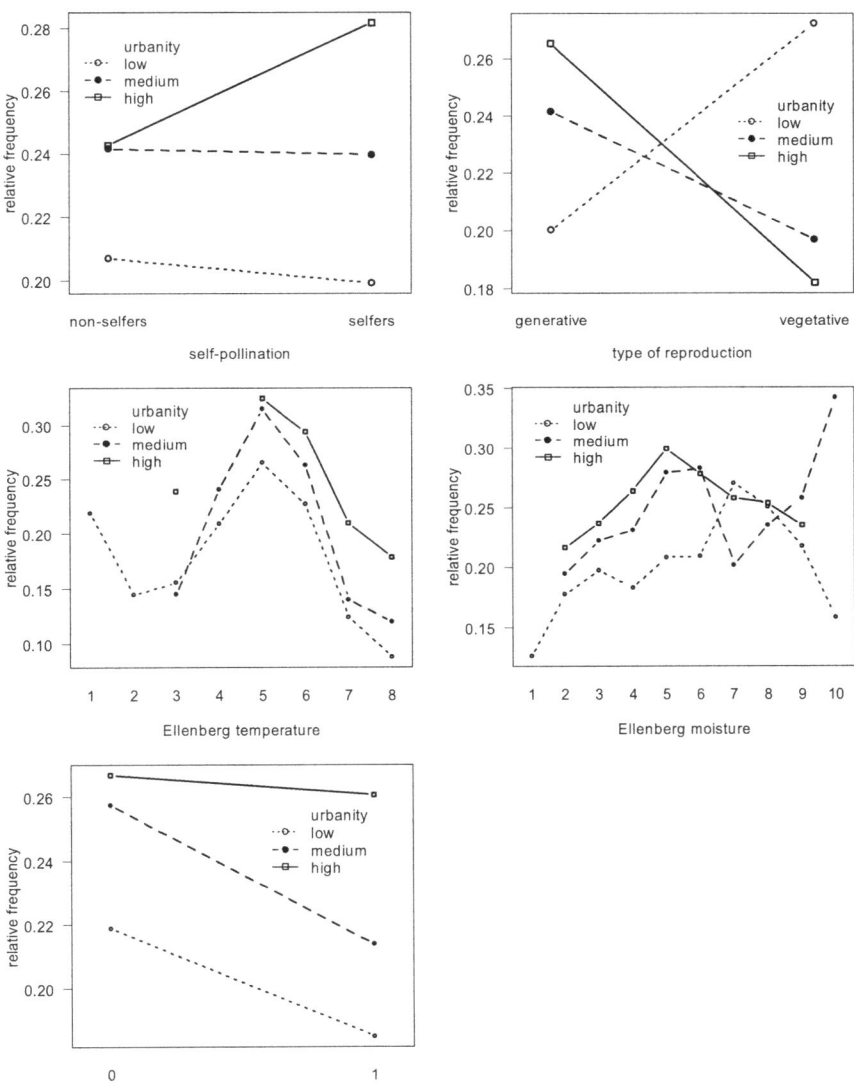

Figure 3. 2 – Interactions between the affinity of plants to urban land use and self-pollination, type of reproduction, Ellenberg temperature, Ellenberg moisture and the existence of hemirosettes

Ellenberg temperature has values ranging from 1 to 8 corresponding to cold to hot habitats (normally values go up to 9, but plants of category 9 did not occur). Ellenberg moisture has values ranging from 1 to 10 corresponding to dry to wet habitats (normally values go up to 12, but plants of categories 11-12 were not used in the analyses). Lines were included for clear illustration, but only when no value in-between two other values was missing. For this figure, urbanity was divided in three classes with low: urbanity $\leq$ 0; medium: $0 <$ urbanity ≤ 0.1; high: $0.1 <$ urbanity ≤ 0.5. Relative frequency was arc-sin transformed.

4.2. Urbanity and Relative Frequencies

Plant species with a high urbanity are relatively frequent, since urbanized areas harbor many common species but only few rare native species (Kühn & Klotz 2006). Many rare species may already have disappeared from urbanized areas. Deleting urbanity and its interactions from the multi-trait model decreased model performance by about one fourth (Table 3.3). Thus, the adaptation to urban land use is clearly important for plant species in Germany, a country with a high settlement density.

4.3. Effects of Trait-Urbanity Interactions on Relative Frequencies

Both selfers and non-selfing species are more frequent with a high than with a low urbanity. However, this pattern is more pronounced for selfers, and species with a high urbanity are more frequent when they are capable of self-pollination (Fig. 3.2a). This confirms what has already been discussed in Chapter I: Many urban habitats are quite young and might have reduced pollinator availability, thus selfing is a suitable strategy for fragmented and disturbed urbanized areas.

The relative frequency of species that reproduce by seeds increases with urbanity (Fig. 3.2b). Many successful urban species have small wind-dispersed seeds that can easily spread in fragmented urban landscapes (Gilbert 1989). Furthermore, urban habitats are subjected to intensive disturbances, e.g. by pedestrians, building activities or the regular mowing of parks (Gilbert 1989; Niemelä 1999). Species with a high dispersal capacity have a higher ability to reach scattered habitat patches (Ozinga et al. 2005) and plants reproducing generatively can settle new habitats faster than plants reproducing vegetatively. Additionally, seeds may persist in spite of disturbances but vegetative propagules may not.

Furthermore, plants are most frequent when preferring moderately warm habitats (Fig. 3.2c), while plants preferring hot habitats are less frequent than plants preferring cold habitats (Korneck et al. 1998; Römermann et al. 2008). However, the higher the urbanity of heat-preferring plants, the more frequent they are, because urban landscapes provide many suitable habitats (e.g. warm railway or industrial sites) and generally have higher temperatures than surrounding rural areas (Oke 1982), hence corresponding to plants preferring dry habitats: The latter as well as species preferring moderately wet habitats are less frequent with decreasing urbanity in four of the multi-trait models (Fig. 3.2d; the high relative frequency of plants with medium urbanity and preferences for wet habitats is only based upon two values and should not be overestimated). Plants preferring wet habitats only find a few suitable habitats in cities (Thompson & McCarthy 2008), while plants preferring dry habitats find many suitable habitats there, reflecting the decrease of humidity along rural-to-urban gradients (Kuttler 2008).

Lastly, species with hemirosettes are more frequent the higher their urbanity (Fig. 3.2e). This is also true for species without hemirosettes, but less pronounced. The number of species with hemirosettes decreased in the countryside due to the abandonement (MacDonald et al. 2000) or intensification of grasslands and pasture lands. Those species with hemirosettes that can cope with ruderal conditions found suitable habitats in cities, e.g. along roadsides where disturbance favors short-lived species but trampling frequency is low (cf. Briemle et al. 2002).

4.4. Applicability and Conclusions

Modelling the potential frequencies of vascular plant species in Germany with environmental niche models implies some uncertainties that might have influenced our results: Within its geographic range (reflected by its actual frequency), a species is affected by a variety of local factors (e.g. biotic interactions such as predation, competition and mutualism; Hampe 2004), which could not be included in our analyses. Small-scale abiotic factors such as disturbance intensity and soil nutrients could also not be included in the niche models. Because small-scale habitat conditions are often crucial for plant performance (Korneck *et al.* 1998) but large-scale climate influences species' ranges (Hampe 2004), future studies should include effects acting on different scales. Our approach of including large-scale climate, topography, soils, and geology as well as the co-occurrence of habitat-specific species reflecting small-scale habitat conditions is a first step in this direction. For instance, habitat-specific species could be proxies for disturbance intensity and soil nutrients.

Many rare species might have already gone extinct due to urbanization and are thus not included in our analyses. Hence, our study shows the current state of the German flora and we cannot make statements about formerly rare but now extinct species, which might add valuable information about the causes of rarity.

Our study showed that it is especially species that depend on other pollinators than themselves, that reproduce vegetatively or that prefer moist, cool or non-ruderal habitats that are threatened by urbanization. Consequently, remnants of semi-natural landscape within urbanized areas, such as alluvial habitats along rivers or forests, might improve living conditions for these species within cities. Moreover, habitats that represent older successional stages than typical urban-industrial habitats might provide better conditions for biologically pollinated species and species that reproduce vegetatively. Accordingly, results might look different when applying this study to cities with a high percentage of semi-natural or older habitats. Vice versa, effects of urban land use on rare species might be even more intensive in countries, which do not have such a long history of human land use as the European countries, and where plants thus had less time to adapt to human land use.

Generally, it is hard to grasp why species are rare. Our results, however, emphasize the need to concentrate on both, species traits and effects of different land-use types to assess species rarity (cf. Fréville *et al.* 2007). Such analyses might enlighten our understanding of rarity and help to derive better conservation strategies, such as creating and protecting habitats that especially support rare species, also within urban areas.

Table 3. 3 – Results of the multi-trait linear models carried out on relative species frequencies in Germany and the relevance of life-history traits and plant affinity to urban land use

See below for further explanation.

	Model R²	URB	Log(CH)	Log(CH):URB	Anemochory	Anemochory:URB	Hemerochory	Hemerochory:URB	Ellenberg_F	Ellenberg_F:URB	Ellenberg_N	Ellenberg_N:URB
∑ positive effects		23	24	0	24	0	24	0	3	0	24	0
∑ negative effects		1	0	0	0	0	0	0	1	4	0	0
All	0.31	+	+		+		+				+	
- URB	0.23		+		+		+				+	
- Log(CH)	0.31	+			+		+				+	
- Anemochory	0.30	+	+				+				+	
- Hemerochory	0.30	+	+		+				-	-	+	
- Ellenberg_F	0.31	+	+		+		+				+	
- Ellenberg_N	0.30	+	+		+		+					
- Ellenberg_T	0.26	-	+		+		+				+	
- Archaeophyte	0.31	+	+		+		+				+	
- Neophyte	0.28	+	+		+		+				+	
- FZ_allrounder	0.31	+	+		+		+				+	
- FZ_temperate-meridional	0.31	+	+		+		+		+	-	+	
- LD_regular	0.31	+	+		+		+				+	
- LD_scarce	0.31	+	+		+		+				+	
- Hemirosettes	0.31	+	+		+		+				+	
- Evergreen leaves	0.31	+	+		+		+				+	
- Summer green	0.31	+	+		+		+				+	
- Hemicryptophytes	0.30	+	+		+		+				+	
- Phanerophytes	0.31	+	+		+		+				+	
- Therophytes	0.31	+	+		+		+				+	
- Insect-pollination	0.31	+	+		+		+				+	
- Self-pollination	0.31	+	+		+		+		+	-	+	
- Wind-pollination	0.31	+	+		+		+				+	
- Log(Seed mass)	0.31	+	+		+		+				+	
- Seed reproduction	0.31	+	+		+		+		+	-	+	

Table 3.3. – continued

	Model R²	Ellenberg_T	Ellenberg_T:URB	Archaeophyte	Archaeophyte:URB	Neophyte	Neophyte:URB	FZ_allrounder	FZ_allrounder:URB	FZ_temperate-merid.	FZ_temp.-merid.:URB	LD_regular	LD_regular:URB
∑ positive effects		0	0	0	0	0	0	24	0	0	0	3	0
∑ negative effects		24	23	1	0	24	0	0	0	24	0	0	0
All	0.31	-	-			-		+		-			
- URB	0.23	-				-		+		-			
- Log(CH)	0.31	-	-			-		+		-		+	
- Anemochory	0.30	-	-			-		+		-			
- Hemerochory	0.30	-	-			-		+		-			
- Ellenberg_F	0.31	-	-			-		+		-			
- Ellenberg_N	0.30	-	-			-		+		-			
- Ellenberg_T	0.26			-		-		+	-	-			
- Archaeophyte	0.31	-	-			-		+		-			
- Neophyte	0.28	-	-					+		-		+	
- FZ_allrounder	0.31	-	-			-				-			
- FZ_temperate-meridional	0.31	-	-			-		+					
- LD_regular	0.31	-	-			-		+		-			
- LD_scarce	0.31	-	-			-		+		-			
- Hemirosettes	0.31	-	-			-		+		-		+	
- Evergreen leaves	0.31	-	-			-		+		-			
- Summer green	0.31	-	-			-		+		-			
- Hemicryptophytes	0.30	-	-			-		+		-			
- Phanerophytes	0.31	-	-			-		+		-			
- Therophytes	0.31	-	-			-		+		-			
- Insect-pollination	0.31	-	-			-		+		-			
- Self-pollination	0.31	-	-			-		+		-			
- Wind-pollination	0.31	-	-			-		+		-			
- Log(Seed mass)	0.31	-	-			-		+		-			
- Seed reproduction	0.31	-	-			-		+	-	-			

Table 3.3. – continued

	Model R^2	LD_scarce	LD_scarce:URB	Hemirosettes	Hemirosettes:URB	Evergreen	Evergreen:URB	Summergreen	Summergreen:URB	Hemicryptophytes	Hemicryptophy.:URB	Phanerophytes	Phanerophytes:URB
∑ positive effects		0	0	0	5	1	1	0	0	24	0	0	0
∑ negative effects		0	0	24	0	1	0	2	0	0	0	2	0
All	0.31			-						+			
- URB	0.23			-				-		+			
- Log(CH)	0.31			-	+					+			
- Anemochory	0.30			-						+			
- Hemerochory	0.30			-						+			
- Ellenberg_F	0.31			-	+					+			
- Ellenberg_N	0.30			-						+		-	
- Ellenberg_T	0.26			-	+	+	+			+			
- Archaeophyte	0.31			-						+			
- Neophyte	0.28			-	+					+			
- FZ_allrounder	0.31			-						+			
- FZ_temperate-meridional	0.31			-						+			
- LD_regular	0.31			-						+			
- LD_scarce	0.31			-						+			
- Hemirosettes	0.31									+		-	
- Evergreen leaves	0.31			-						+			
- Summer green	0.31			-						+			
- Hemicryptophytes	0.30			-		-							
- Phanerophytes	0.31			-						+			
- Therophytes	0.31			-						+			
- Insect-pollination	0.31			-						+			
- Self-pollination	0.31			-	+					+			
- Wind-pollination	0.31			-						+			
- Log(Seed mass)	0.31			-						+			
- Seed reproduction	0.31			-				-		+			

Table 3.3. – continued

	Model R^2	Therophytes	Therophytes:URB	Insect-pollination	Insect-pollination:URB	Self-pollination	Self-pollination:URB	Wind-pollination	Wind-pollination: URB	Log(Seed mass)	Log(Seed mass):URB	Seed reproduction	Seed reproduction:URB
∑ positive effects		3	0	0	0	0	23	0	0	0	0	0	23
∑ negative effects		0	0	0	0	23	0	0	0	0	0	23	0
All	0.31					-	+					-	+
- URB	0.23	+											
- Log(CH)	0.31					-	+					-	+
- Anemochory	0.30					-	+					-	+
- Hemerochory	0.30	+				-	+					-	+
- Ellenberg_F	0.31					-	+					-	+
- Ellenberg_N	0.30	+				-	+					-	+
- Ellenberg_T	0.26					-	+					-	+
- Archaeophyte	0.31					-	+					-	+
- Neophyte	0.28					-	+					-	+
- FZ_allrounder	0.31					-	+					-	+
- FZ_temperate-meridional	0.31					-	+					-	+
- LD_regular	0.31					-	+					-	+
- LD_scarce	0.31					-	+					-	+
- Hemirosettes	0.31					-	+					-	+
- Evergreen leaves	0.31					-	+					-	+
- Summer green	0.31					-	+					-	+
- Hemicryptophytes	0.30					-	+					-	+
- Phanerophytes	0.31					-	+					-	+
- Therophytes	0.31	░	░			-	+					-	+
- Insect-pollination	0.31			░	░	-	+					-	+
- Self-pollination	0.31					░	░					-	+
- Wind-pollination	0.31					-	+	░	░			-	+
- Log(Seed mass)	0.31					-	+			░	░	-	+
- Seed reproduction	0.31					-	+					░	░

Further explanation for table 3.3.: Given are importance and effects of traits or trait-states and trait-urbanity interactions (URB = urbanity) for the relative frequency of plant species. Each predictor was excluded once (e.g. -Neophyte: interaction neophyte:URB and the trait neophyte without interaction were excluded; the according cells are shaded in grey). '+' and '-' show whether a predictor was present in the minimal adequate model, with '+' indicating a positive effect and '-' indicating a negative effect on relative species frequency. Model R^2 shows the R^2 adjusted for the number of predictors for each model. CH = Canopy height

Table 3. 4 – Results of the final model carried out on relative species frequencies in Germany and the relevance of life-history traits and plant affinity to urban land use
The final model ($R^2 = 0.31$***) included all predictors that occurred in 23 or 24 of the 25 multi-trait models. Estimates indicate whether the traits, urbanity (URB) and trait-urbanity interactions positively (+) or negatively (-) affected relative species frequency and show the strength of the slopes and the model intercept. Predictors were standardized to zero mean and unit standard deviance. Levels of significance: *** $P \leq 0.001$, * $P \leq 0.05$, + $P \leq 0.1$, n.s. not significant.

Trait / URB	Estimate	p-value
Intercept	0.25	***
Urbanity	0.04	***
Log(Canopy height)	0.01	+
Anemochory	0.01	***
Hemerochory	0.02	***
Hemirosettes	-0.02	***
Neophyte	-0.02	***
FZ_allrounder	0.01	*
FZ_temperate-meridional	-0.01	*
Hemicryptophytes	0.02	***
Ellenberg_N	0.01	***
Self-pollination	0.001	n.s.
Self-pollination:URB	0.007	*
Seed reproduction	-0.001	n.s.
Seed reproduction:URB	0.007	*
Ellenberg_T	-0.04	***
Ellenberg_T:URB	-0.03	***

Chapter IV – Changes in the Functional Composition of a Central European Urban Flora over Three Centuries

1. Introduction

Urbanization has shaped European landscapes for many centuries. The first towns already developed around 700 B.C. in the Mediterranean (Antrop 2004). Since these early times, urbanization spread all over Europe which is today one of the most urbanized continents, with 72% of the total population living in urban areas (only Latin and Northern America have higher rates of urban population with 78% and 81% respectively; United Nations 2008). In the 18th century and especially in the 19th century, industrialization and trade caused the growth of many European towns (Berry 1990). However, the main phase of urbanization took place in the 20th century (Berry 1990; United Nations 2006) with its rapid developments in transportation techniques (Berry 1990; Antrop 2004). The increased mobility, together with other factors, such as political frameworks, enabled urban sprawl, which was especially strong in the second half of the 20th century (Kasanko *et al.* 2006).

Urbanization changes landscapes profoundly. In Europe, land use often changed from agricultural to urban but also from semi-natural to urban (Kasanko *et al.* 2006). These changes have severe impacts on climate, biogeochemical cycles, hydrology and biodiversity (Vitousek *et al.* 1997): Compared to rural surroundings, the high heat capacity of buildings together with heating increases urban temperatures (Landsberg 1981; Oke 1982; Sukopp 1998); the emission of pollutants from traffic, industries and heating changes the composition of the atmosphere (Berry 1990); decomposition rates and nitrification rates increase in urban relative to rural forest stands (McDonnell *et al.* 1997); the high proportion of sealed surfaces reduces infiltration capacity and groundwater replenishment

(Sukopp 1998); proportions of native species decrease, while proportions of non-native species increase (Kowarik 2008).

It is clear from the characteristics of urban environments that not every species is able to persist there. Indeed, species with adaptations to disturbance, fragmentation, high temperature, or drought, i.e. species with traits that enable them to cope with urban conditions, are more frequent in cities than in the countryside (Wittig & Durwen 1982; Lososová *et al.* 2006; Chapters I and II). Many of such differences in the composition of trait states of urban and rural species assemblages have been shown in space, and other studies showed that trait state composition also changes over time (e.g. Chocholoušková & Pyšek 2003; Pyšek *et al.* 2004; Van der Veken *et al.* 2004; Tait *et al.* 2005; Tamis *et al.* 2005; Lavergne *et al.* 2006).

We studied the development of the flora in the city of Halle in Central Germany over 320 years. The earliest available relatively complete floristic records for Halle date back to the year 1687, the most recent records were published in 2004. For the three centuries in-between, several floristic mappings are available, covering nearly the whole time-span. We are not aware of many other databases on terrestrial plants covering such a long time-span (but see Preston 2000 for Cambridgeshire and Middlesex, UK). This gives us a unique opportunity to study changes in plant species assemblages exposed to more than 300 years of urbanization. We assume that changes over time reflect differences between urban and rural areas in space, with species adapted to urban characteristics increasing their proportion in the flora as urbanization intensifies.

2. Materials and Methods

2.1. Study Area

2.1.1. Population Development

Shortly after the first time period of our analysis (1687-1689, see Table 4.1), the formation of the university in 1694 and of the "Franckesche Stiftungen" (school and orphanage with international importance) from 1698 to 1745 gave rise to a significant increase in population numbers (reaching around 21,000 in 1820; Stolle & Klotz 2004). Industrialization caused a second period of growth between 1850 and 1900 (Walossek 2006; corresponding to the fourth time period of our analysis: 1857-1901). In 1900, the population reached 156,636 (this and the following population figures from http://www.halle.de).

From 1924 to 1930, population again rose significantly and new districts were built (Walossek 2006). After World War II, population numbers reached 293,113 (in 1952). Halle now belonged to the GDR and between 1964 and the 1980ies, a completely new city was built, Halle-Neustadt, just at the other side of the river Saale, opposite to the old town. Together, Halle and Halle-Neustadt had 309,406 inhabitants after the German reunification in 1990. The critical economic situation in Eastern Germany after the reunification led to a drastic decrease in population numbers (Raschke & Schultz 2006). Today (2008), the administrative district of Halle, now consisting of both Halle and Halle-Neustadt, has 231,778 inhabitants.

2.1.2. Environmental Conditions

For a long time, the river Saale served as the city's discharge system for untreated wastewater. It was not until 1915 that a sewage treatment plant was established. However, the river's contamination continued, especially in the GDR (1949-1990), when wastewater from the chemical industry south of Halle – the largest plants of the chemical industry in the GDR – but also from lignite and potash mining was dumped in the river (Walossek et al. 2006). Habitats for aquatic plants were significantly reduced: In the 1970ies and 1980ies, no higher plants were able to live in the river (Stolle & Klotz 2004). Similarly, the air was contaminated by soot, ashes and sulfur dioxide from industry and heating with fuelwood and lignite (Zierdt 2006).

With the German reunification, many industries were closed down or modernized step-by-step, as were the heating systems of private households. Both water and air quality improved significantly since then (Walossek *et al.* 2006; Zierdt 2006). However, atmospheric nutrient inputs still affect vegetation.

2.2. Data Sources

2.2.1. Species Data

We analyzed several recent and historical floras as well as smaller manuscripts with descriptions of plant occurrences from more than 18 authors and divided the data into seven time periods (see Table 4.1 for overview and references). Regarding species occurrences, we always referred to the current (2004) administrative district of Halle and not to the administrative district of the respective time periods. Therefore, the study area today includes the city of Halle, but included the city and its immediate rural vicinity in former time periods, when the city had not yet reached today's extent.

We are aware of the fact that the floristic investigations of the 17th to 19th century were not as exhaustive as today's investigations. However, with 820 plant species occurring in the first time period (1687-1689), these early investigations already covered 82% of the species occurring in the last time period (2000-2008 with 1000 species; Table 4.2). Thus, the data are well representative and allow meaningful statistical analyses.

To exclude as many uncertainties as possible, we assessed the historical floras for the plausibility of species occurrences. We excluded problematic species, i.e. species which are unlikely to have grown in the study area or species that could not be assigned to modern nomenclature. Species that were highly likely to have occurred within the study area itself but which were only listed with occurrences in the surroundings of the study area and were therefore probably overlooked by former botanists, were included as occurring in the study area. Species occurring exclusively in cultivation were excluded.

Species which we distinguish nowadays but that were not distinguished in the past were lumped together as so-called superspecies. Therefore, we took the superspecies from the first period (1687-1689) as basis and assigned all associated species to this superspecies for all periods, also when the single species were distinguished in the following periods.

Table 4. 1 – Overview of the time periods used in the analyses on the historical flora of Halle

Shown are the length of each period, the mean of each period used as predictor in linear models (see *Data analyses* section of this Chapter) and respective reference floras

Length of time period	Mean of time period	Reference
1687 - 1689	1688	Knauth 1687; improved edition 1689
1721 - 1783	1752	Buxbaum 1721; Senckenberg 1731, published in Spilger 1937; Leysser 1761; 1783; Roth 1783
1806 - 1856	1831	Luyken 1806 (not published but documented in the herbarium of the Westphalian Museum of Natural History in Münster); Sprengel 1806; Wallroth 1815; 1822; Garcke 1848; 1856
1857 - 1901	1879	Fitting *et al.* 1899; 1901; herbarium of the University of Halle
1902 - 1949	1925	Fitting *et al.* 1903; Schulz & Wüst 1906; 1907; Wangerin 1909; Knapp 1944a; 1944b; 1945; several unpublished manuscripts by M. Schulze (1936, 1938) stored in the archive of the working group of hercynian florists; herbarium of the University of Halle
1950 - 1999	1975	Rauschert 1966a; 1966b; 1967; 1972; 1973; 1975; 1977a; 1977b; 1979; 1980; 1982 and several unpublished manuscripts by S. Rauschert (1959-1982); Grosse 1978; 1979; 1981; 1983; 1985; 1987; Klotz 1984; Grosse & John 1987; 1989; 1991; Klotz & Stolle 1998; herbarium of the University of Halle
2000 - 2008	2004	Stolle & Klotz 2004; unpublished data by J. Stolle and S. Klotz (2005 – 2008)

Figure 4. 1 – Urban development of the city of Halle, 1740-2002
The city of Halle in (a) 1740, (b) 1942/43 and (c) 2002. The city area of (a) is indicated in (b) and (c) by a black circle. References: (a) Historischer Stadtplan Halle a. d. Saale. Kol. Kupferstich anno 1740, vermutlich von Johann David Schleuen (Verlag) Berlin. Nachdruck. Bildarchiv Preußischer Kulturbesitz. Original in der Staatsbibliothek zu Berlin, Kartenabteilung. Horst Hup Edition Berlin II/2004; (b) Topographische Karten 1: 25 000 (TK 4437, 4438, 4537, 4538), Preußische Landesaufnahme (Hrsg.) 1904 (TK 4437) / 1905 (TK 4438, 4537, 4538), Reichsamt für Landesaufnahme, Ausgabe 1942 (TK 4437, 4538)/ 1943 (TK 4438, 4537); (c) Topographische Karte 1: 100 000, Regionalkarte Sachsen-Anhalt, Raum Halle-Merseburg, Harzvorland (TK 100 RK/Blatt 4). Landesamt für Landesvermessung und Datenverarbeitung Sachsen-Anhalt 2002. Halle (Saale)

This procedure minimized inconsistencies and false occurrences as well as pseudo-absences in the floras; however, if a species was overlooked by former botanists, we do not have any evidence for its occurrence and cannot reconstruct it. This concerns native species and archaeophytes but does not apply to neophytes, because most naturalizations of neophytes in Germany took place after 1850 (Kühn & Klotz 2002; for Halle see Fig. 4.2) and therefore well after the first two time periods (1687 – 1689, 1721 – 1783), which have the highest probability of being inconsistent. If a species is not naturalized in Germany, it is of course not naturalized in Halle as well. To further account for the potential incompleteness of early floras, we performed our analyses once with the first time period as basis and a second time with the third time period as basis. Plant mapping in the third period is mainly based on Garcke (1848; 1856), whose flora is not only the most complete of the early floras but also revised faults of preceding floras (see Table 4.2 for species numbers per time period).

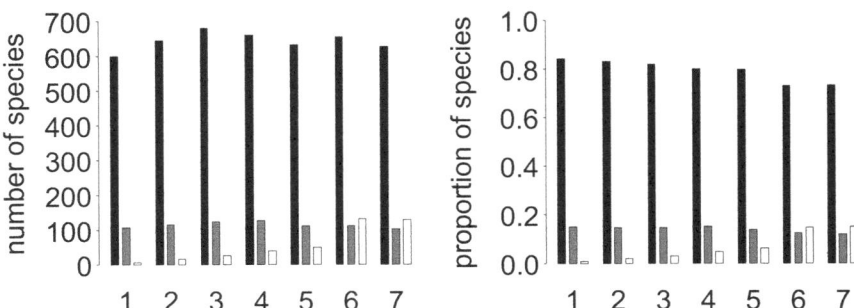

Figure 4. 2 – Number and proportion of species per type of floristic status in each of the seven time periods used for the historical analyses of the flora of Halle
Left: Number; Right: proportion of species. Time periods correspond to 1: 1687-1689; 2: 1721-1783; 3: 1806-1856; 4: 1857-1901; 5: 1902-1949; 6: 1950-1999; 7: 2000-2008. Black: native species; grey: archaeophytes; white: neophytes.

2.2.2. Trait Data

As for the preceding chapters, trait data were taken from BiolFlor (Klotz *et al.* 2002; Kühn *et al.* 2004b; http://www.ufz.de/biolflor), LEDA (Kleyer *et al.* 2008; http://www.leda-traitbase.org), and additionally from Ellenberg *et al.* (2001). We chose traits (Table A1) related to dispersal, persistence, phenology, pollination, and the plants' preferences for local moisture regime, temperature regime and soil's nitrogen content (Ellenberg-values). The floristic status of a species (whether it is native, archaeophyte or neophyte in Germany as recorded in BiolFlor) was used to distinguish between extinctions and introductions.

Table 4. 2 – Number (#) of plant species, extinct native species, extinct archaeophytes, and introduced neophytes in the Halle region per time period
Numbers of extinctions and introductions are not shown for 1687-1689, because this is the reference period (see *Materials and Methods* section of this Chapter for details). Differences in the number of species between the periods deviate from the sum of extinctions and introductions because not only the neophytes shown here were introduced but also natives and archaeophytes (not shown).

Time period	# plant species	# extinct natives	# extinct archaeophytes	# introduced neophytes
1687 – 1689	820	-	-	-
1721 – 1783	904	5	-	10
1806 – 1856	966	13	1	12
1857 – 1901	956	31	2	14
1902 – 1949	918	50	13	14
1950 – 1999	1045	27	3	83
2000 – 2008	1000	19	4	1

2.3. Data Analyses

2.3.1 Association of Trait States with Time

We first calculated the number of species per trait state and time period. χ^2-tests or Fisher's exact tests, respectively, showed whether trait states were associated with specific time spans. Fisher's exact test had to be used if the number of at least one of the expected elements per time period was < 5 (Crawley 2007). Because changes in trait state numbers might become clearer when only comparing the first and last time period, we repeated the analyses for 1687-1689 and 2000-2008.

2.3.2 Association of Trait States with Extinction and Introduction

Additionally, we calculated the number of species per trait state for (i) all native plant species and (ii) all archaeophytes that were extinct after 1689, and (iii) for all neophytes that were introduced since 1689. We tested for associations of trait states with extinction or introduction by comparing species numbers of trait states in the basis period (1687-1689) with the species numbers of (i), (ii) and (iii) using χ^2-test or Fisher's exact test, respectively. To exclude patterns that might occur because of incomplete mapping in 1687-1689, we repeated these calculations with the third time period (1806-1856) as a reference period.

Table 4. 3 – β_{sim}-similarity index comparing the presence and absence of plant species between pairs of time periods for the flora of Halle

β_{sim} is calculated as $\beta_{sim} = a / (a + min (b, c))$, where a is the number of species shared between two periods and b and c are the numbers of species unique to a period. Values range from zero to one, with the upper limit indicating complete similarity and the lower limit indicating no similarity between the species lists of two periods.

	1687-1689	1721 – 1783	1806 – 1856	1857 – 1901	1902 – 1949	1950 – 1999
1687-1689						
1721 - 1783	0.97					
1806 - 1856	0.96	0.94				
1857 - 1901	0.93	0.90	0.93			
1902 - 1949	0.84	0.81	0.88	0.92		
1950 - 1999	0.81	0.80	0.81	0.84	0.95	
2000 - 2008	0.78	0.76	0.77	0.80	0.91	0.99

2.3.3. Trends in Trait State Ratio Development

χ^2-test and Fisher's exact test are based on species numbers, not on species proportions per time span. However, the proportion of one trait state to another gives more detailed information about compositional changes of the flora. Because the different states of a trait add up to 100%, we calculated log-ratios of proportions to break this unit sum constraint (see Chapters I and II; Aitchison 1982; Elston *et al.* 1996; Billheimer *et al.* 2001; Kühn *et al.* 2006).

First, we tested the log-ratios for temporal autocorrelation. Temporal autocorrelation occurs because the flora of a time period partly depends on the flora of the preceding time period, i.e. the species pool (Zobel 1997) of period A is the basis for the species pool of period B. This is illustrated by species turnover, which increases with increasing temporal distance between periods (Table 4.3). However, none of the log-ratios in our analysis was significantly autocorrelated in time (not shown); thus, there was no need to account for temporal autocorrelation in the linear models Therefore, it was possible to use simple linear models with the log-ratios as response and the mean of each time period as predictor. We tested whether there is a significant positive or negative trend for the development of trait state ratios over time (Crawley 2007). Here, we considered the total flora per time period, not only extinct or introduced species. We used the mean of each time period (e.g. 1688 for 1678-1689 or 1925 for 1902-1949; see Table 4.1) to account for the different lengths of time-lags in-between the periods.

All calculations were performed with the open source software R, version 2.6.0 (R Development Core Team 2007).

3. Results

3.1. Association of Trait States with Time

None of the traits but life form ($p < 0.05$) and floristic status ($p < 0.001$) was associated with time when comparing species numbers per trait state for all time periods.

Comparing only the first and the last time period showed changes for Ellenberg moisture ($p < 0.05$), Ellenberg nitrogen ($p < 0.05$), floristic status ($p < 0.001$), leaf anatomy ($p < 0.05$) and life form ($p < 0.01$).

Species of dry to fresh soils, species of inundated soils and aquatic species increased in numbers between 1687-1689 and 2000-2008, while species of moist to wet soils (among them many species growing in bogs) decreased. Species of nitrogen-poor habitats decreased, and species preferring medium nitrogen contents or nitrogen-rich habitats increased. The number of native species and archaeophytes decreased, while the number of neophytes increased. Species with helomorphic leaves decreased their numbers, and species with mesomorphic or hygromorphic leaves increased their numbers. The number of hemicryptophytes decreased, while the number of phanerophytes increased.

3.2. Association of Trait States with Extinction and Introduction

Using the first time period as basis for comparison with extinct and introduced species mostly yielded the same results as using the third time period. Therefore, the following results apply to both calculations, except where indicated otherwise. Numbers of extinct and introduced species per time period are shown in Table 4.2.

Dispersal type was associated with introduction but not with extinction: Introduced species were more often dispersed by humans but less often dispersed by wind than expected in the reference period (Table 4.4, illustrated with proportions in Fig.4.3a). Species preferences towards their habitat's moisture regime were associated with the extinction of native species, but not with the extinction of archaeophytes or the introduction of neophytes: Native species of moist to wet soils, e.g. plants growing in bogs, got preferably extinct, while species preferring drier soils were less often extinct than expected (Table 4.4, Fig.4.3b). Native species preferably growing in nitrogen-poor soils were marginally associated with extinction, but only when using the third time period as basis for comparison. Furthermore, neophytes were introduced less often than expected when preferring low nitrogen contents, but more often when preferring high nitrogen contents (Table 4.4, Fig.4.3c). Species preferences towards their habitat's temperature regime were associated with extinction and introduction, with native species of (moderately) cool habitats and natives and archaeophytes of warm habitats becoming extinct more often than expected. However, neophytic species preferring warm habitats were also introduced more often than expected (Table 4.4, Fig.4.3d-f). Flowering phenology was neither associated with extinction, nor with introduction (Table 4.4). Native species with hydromorphic or helomorphic leaves got preferably extinct, while neophytes with these two trait states were introduced less often than expected. Moreover, natives with hygromorphic or mesomorphic leaves got less often extinct than expected and introduced neophytes were most often mesomorphic (Table 4.4, Fig.4.3g-h). Concerning leaf persistence, none of the native species extinct after 1689 was spring-green and also plants with overwintering leaves got extinct less often than expected. Contrarily, summer green leaves were over-represented among extinct natives. Leaf persistence was only marginally associated with introduction when using the first time period as basis for comparison, with species having evergreen or overwintering green leaves being introduced less often but species having summer green leaves being introduced more often than expected (Table 4.4, Fig.4.3i-j). Phanerophytes and therophytes were over-represented among introduced neophytes, while other life forms, such as hydrophytes or hemicryptophytes were under-represented (Table 4.4, Fig.4.3k). Life span showed no association with extinction and introduction (Table 4.4). Insect-pollinated species were introduced more often and wind-pollinated species less often than expected (Table 4.4, Fig.4.3l). Lastly, specific leaf area was neither associated with extinction nor with introduction (Table 4.4).

3.3. Trends in Trait State Ratio Development

The trend analyses mainly confirmed the tests for association with extinction and introduction (see Table 4.5 for all trends of trait state ratios). Additionally, the proportions of animal-dispersed species increased from 1687 to 2008, at the expense of wind-dispersed species. Species of inundated soils and aquatic species increased their proportion at the expense of species preferring (extremely) dry soils or moist to wet soils. Neophytes increased their proportion significantly, at the expense of natives and archaeophytes. Species that flower in prespring were the only phenological group whose proportion did not decrease when compared with the increase of species flowering in early autumn. The proportions of therophytes decreased, but only relative to phanerophytes, which increased their proportions in the flora of Halle. Besides insect-pollinated species, also self-pollinated species increased their proportions at the expense of wind-pollinated species.

4. Discussion

Historical data of course are associated with uncertainties. Incomplete mappings that can not be completed several hundred years later can give a wrong picture of former floras. Our dataset, however, is reliable enough to give representative results due to several reasons: First, botany has a long tradition in the Halle region, where Johannes Thal in 1577 wrote the first known flora worldwide (Rauschert 1977a) that did not only concentrate on pharmaceutical or agricultural plants but included as many plant species as possible and covered a complete region (Thal 1588, published in reprint in 1977). Even Carl v. Linneé appreciated Thal's work and named *Arabidopsis thaliana* (L.) HEYNH. after him. Many others followed, building on Thal's knowledge and passing it to others, like Friedrich Wilhelm v. Leysser who was the first in Germany to consistently use the modern nomenclature of Linnée. Secondly, with 820 species in 1687-1689 and 1000 species in 2000-2008, the representativeness of the earliest flora is already high. Thirdly, we minimized uncertainties by checking the dataset for unlikely occurrences (two of us, J. Stolle and S. Klotz, are experienced botanists who have studied the flora of Halle intensely for many years), and by taking natives and archaeophytes mapped in the first (resp. third) time period as basis. Comparing all species occurring in the reference time period with all species occurring today increases the influence of pseudo-absences. Looking only at those natives and archaeophytes that disappeared since the reference time period solved the problem of pseudo-absences. Fourthly, the flora of the third time period is known to be thoroughly mapped and taking this period as basis for comparison mostly yielded the same results as taking the first time period as basis. Lastly, our study revealed some self-explanatory patterns which are confirmed by other studies, such as the increase of neophytes and the relative decrease of natives and archaeophytes (Godefroid 2001; Chocholoušková & Pyšek 2003; Tait *et al.* 2005), the increase of human-dispersed species and the increase of species of nitrogen-rich habitats (Preston 2000; Godefroid 2001; Tamis *et al.* 2005; Römermann *et al.* 2008).

Urbanization has changed the face of the study area considerably. Halle transformed from a small town in the borders of a city wall in the 17th century to a modern city that forms, together with the neighboring city of Leipzig, the tenth largest conurbation in Germany (Friedrich 2006). The process of urbanization caused a species turnover of 22% in 320 years. Consequently, the changes in the proportions of trait states that we observed in our analysis are likely to reflect the change in land use and the accompanying environmental changes. Some of the developments in the functional composition of the flora also took place in non-urban regions and therefore are indicators of general land-use and climate change: Nitrogen inputs in Europe increased drastically since the advent of modern agricultural techniques, industries and traffic (Franzaring & Fangmeier 2006), leading to a general increase in the proportion of plant species that can grow in nitrogen-rich habitats and a reduction of habitats for plant species that prefer nitrogen-poor conditions (Tamis *et al.* 2005; Römermann *et al.* 2008). In the study area, nitrogen-containing mineral fertilizers were used earlier than elsewhere. The region is characterized by chernozems – the agriculturally most valuable soils in Germany. Therefore, agriculture and the demand for mineral fertilizers were strong.

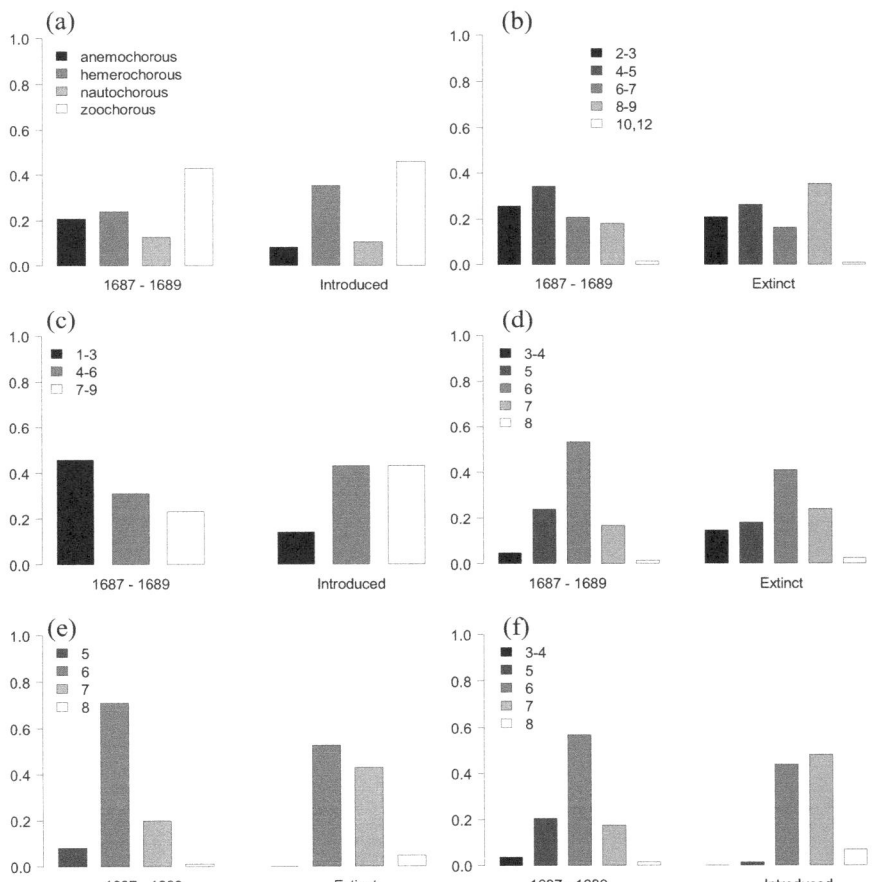

Figure 4. 3 – Association of plant trait states with extinction and introduction in the flora of Halle

Shown is the proportion of each trait state (0.0 – 1.0) in the flora of the basis period (1687-1689) and the proportion of these trait states for either all natives extinct after 1689, all archaeophytes extinct after 1689, or all neophytes introduced after 1689 (selected results; indicated for each trait as follows): (a) dispersal type, neophytes; (b) Ellenberg moisture, natives; (c) Ellenberg nitrogen, neophytes; (d) Ellenberg temperature, natives; (e) Ellenberg temperature, archaeophytes; (f) Ellenberg temperature, neophytes;

(g) leaf anatomy, natives; (h) leaf anatomy, neophytes; (i) leaf persistence, natives; (j) leaf persistence, neophytes; (k) life form, neophytes; (l) pollination type, neophytes.

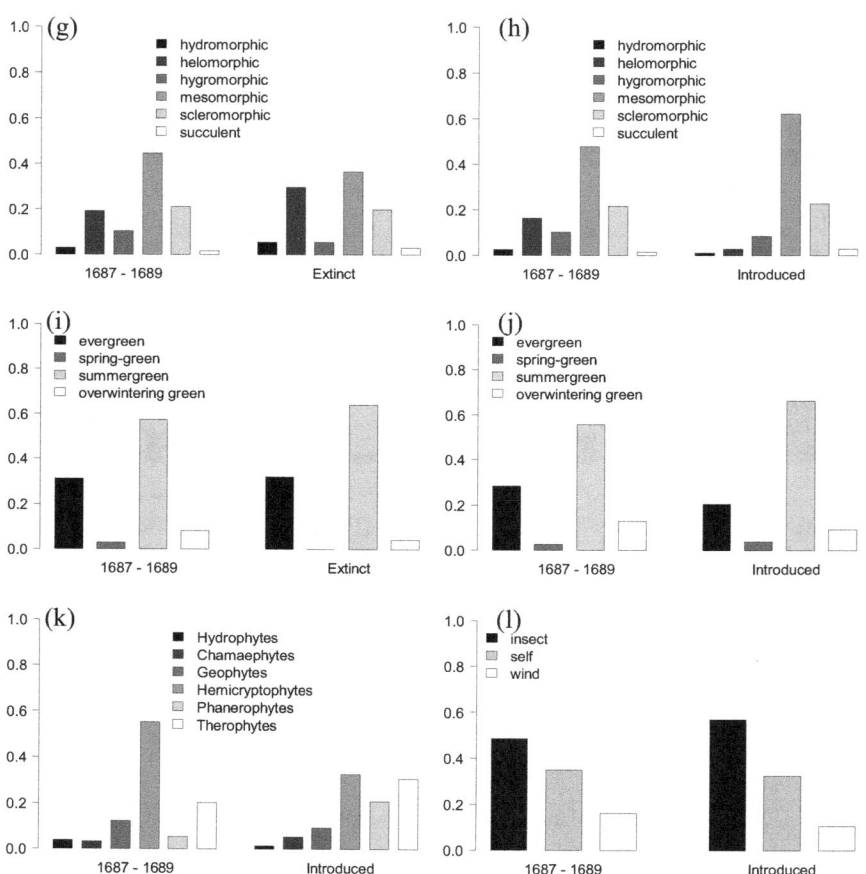

Figure 4.3. – continued

Similar to nitrogen-preferring species, the increase in the proportion of plant species preferring warm habitats is not restricted to urbanized areas (Tamis *et al.* 2005). Urban temperatures exceed temperatures of non-urban areas (Landsberg 1981; Oke 1982; Sukopp 1998), but climate change has increased temperatures of both urban and rural areas (Intergovernmental Panel on Climate Change (IPCC) 2007; http://www.ipcc.ch). However, many neophytes with an origin in warmer climates perform especially well in cities because urban winter temperatures allow them to persist there (Sukopp *et al.* 1979). Thus, the increase of heat preferring plants might be especially high in urban areas anticipating potential developments under climate warming.

Table 4. 4 – Association of plant trait states with extinct natives, extinct archaeophytes, and introduced neophytes in the flora of Halle

Significance of the association with extinct natives ($P_{\text{extinct natives}}$); extinct archaeophytes ($P_{\text{extinct archaeophytes}}$) and introduced neophytes ($P_{\text{introduced neophytes}}$); P-values originate from χ^2-test or Fisher's exact test, respectively (see *Materials and Methods* section of this Chapter for details). Shown is the association in comparison with (i) native species present in 1687-1689 (for natives), (ii) archaeophytes present in 1687-1689 (for archaeophytes) and (iii) natives, archaeophytes and neophytes present in 1687-1689 (for neophytes). $P \geq 0.1$ n.s., $0.1 > P > 0.05+$; $P \leq 0.05*$; $P \leq 0.01**$, $P \leq 0.001***$.

Traits	$P_{\text{extinct natives}}$	$P_{\text{extinct archaeophytes}}$	$P_{\text{introduced neophytes}}$
Dispersal type	> 0.1 n.s	> 0.1 n.s	< 0.01**
Ellenberg moisture	< 0.01**	> 0.1 n.s	> 0.1 n.s
Ellenberg nitrogen	> 0.1 n.s	> 0.1 n.s	< 0.001***
Ellenberg temperature	< 0.01**	< 0.1 +	< 0.001***
Flowering phenology	> 0.1 n.s	> 0.1 n.s	> 0.1 n.s
Leaf anatomy	< 0.01**	> 0.1 n.s	< 0.001***
Leaf persistence	< 0.05*	> 0.1 n.s	< 0.1 +
Life form	> 0.1 n.s	> 0.1 n.s	< 0.001***
Life span	> 0.1 n.s	> 0.1 n.s	> 0.1 n.s
Pollination type	> 0.1 n.s	> 0.1 n.s	< 0.1 +
Specific leaf area	> 0.1 n.s	> 0.1 n.s	> 0.1 n.s

Table 4. 5 – Trends of trait state ratios over the period of the historical analyses for the flora of Halle

Trait state ratios are log-transformed (see *Materials and Methods* section of this Chapter for details). Shown are parameter estimates of linear trend models with their *P*-values and model R^2 (adjusted for number of predictors) with its *P*-value: $P \geq 0.1$ n.s., $0.1 > P > 0.05+$; $P \leq 0.05^*$; $P \leq 0.01^{**}$, $P \leq 0.001^{***}$.

Trait	Log-ratio	Estimate	R^2_{model}
Dispersal type	Hemerochory / Anemochory	$6e^{-4}$***	0.89***
	Hydrochory / Anemochory	$-7.6e^{-6}$ n.s	-0.2 n.s
	Zoochory / Anemochory	$3.9e^{-4}$***	0.9***
Ellenberg moisture	2-3 / 10-12	-0.001*	0.56*
	4-5 / 10-12	-0.0005 n.s	0.16 n.s
	6-7 / 10-12	-0.0007 n.s	0.29 n.s
	8-9 / 10-12	-0.0027**	0.83**
Ellenberg nitrogen	1-3 / 7-9	-0.001***	0.9***
	4-6 / 7-9	$-3.36e^{-7}$ n.s	-0.2 n.s
Ellenberg temperature	3-4 / 8	-0.003**	0.79**
	5 / 8	-0.0007 n.s	0.19 n.s
	6 / 8	-0.001	0.61*
	7 / 8	-0.00054 n.s	0.26 n.s
Floristic status	Archaeophytes / Neophytes	-0.009***	0.96***
	Natives / Neophytes	-0.009***	0.97***
Flowering phenology	Pre-spring / Early autumn	0.0001 n.s	-0.16 n.s
	Early spring / Early autumn	-0.0008*	0.57*
	Mid spring / Early autumn	-0.001**	0.79**
	Early summer / Early autumn	-0.001*	0.62*
	Midsummer / Early autumn	-0.001*	0.64*
Leaf anatomy	Helomorphic / Hygromorphic	-0.002**	0.82**
	Hydromorphic / Hygromorphic	-0.001 +	0.38 +
	Mesomorphic / Hygromorphic	-0.0003 n.s	0.18 n.s
	Skleromorphic / Hygromorphic	-0.0007*	0.68*
	Succulent / Hygromorphic	-0.001 +	0.46
Leaf persistence	Evergreen / Overwintering green	0.0002 n.s	-0.02 n.s
	Spring green / Overwintering green	0.001*	0.57*
	Summer green / Overwintering green	0.0002 n.s	-0.04 n.s

Table 4.5 – continued

Trait	Log-ratio	Estimate	R^2_{model}
Life form	Hydrophytes / Therophytes	-0.0003 n.s	0.004 n.s
	Chamaephytes / Therophytes	0.0004 n.s	0.04
	Geophytes / Therophytes	$6e^{-5}$ n.s	-0.19 n.s
	Hemicryptophytes / Therophytes	-0.0002 n.s	-0.06 n.s
	Phanerophytes / Therophytes	0.002*	0.62*
Life span	Annuals / Biennials	-0.0002 n.s	-0.06 n.s
	Pluriennials / Biennials	$-7e^{-5}$ n.s	-0.17 n.s
Pollination type	Insect / Wind	0.0003 +	0.46 +
	Self / Wind	$2e^{-4}$**	0.74**
Specific leaf area [mm²/mg]	SLA< 20 / SLA> 60	$4e^{-5}$ n.s	-0.2 n.s
	20 < SLA < 40 / SLA > 60	$9e^{-5}$ n.s	-0.2 n.s
	40 < SLA < 60 / SLA > 60	0.0002 n.s	-0.17 n.s

Changes in temperature also cause phenological changes (Roetzer *et al.* 2000; Badeck *et al.* 2004; Lu *et al.* 2006; Luo *et al.* 2007). Flowering as well as leaf phenology in the study area showed changes at both the beginning and end of the vegetation period, indicating an increase of temperature in the Halle region over the last three centuries, probably due to both urbanization and climate change: Flowering starts earlier and stops later, and proportions of plant species with spring green leaves increased.

Phenological changes might as well be the result of land-use changes: Many plants with overwintering green leaves are annual weeds of arable land that flower and reproduce when old crops have been harvested and new crops not yet been sown or are not grown tall. Their proportion in the total flora probably decreased due to the decrease and intensification of agricultural land use.

Besides, gardeners' preferences might influence phenological patterns: Plants flowering in pre-spring (such as *Galanthus nivalis* L.) or autumn (such as *Solidago canadensis* L.) are attractive for gardening, because many gardeners want their gardens to be green and flowering as long as possible. Similarly, insect-pollinated plants often have showy flowers and are thus more attractive than many wind-pollinated species.

Shrubs and trees (phanerophytes) are planted along roads or in parks, among them also neophytes such as *Ailanthus altissima* (Mill.) Swingle or *Robinia pseudoacacia* L. (Kowarik & Säumel 2007), increasing species numbers of phanerophytes in the urban flora.

Because many neophytes reach their 'new home' with trade and traffic, it is a logical consequence that the proportion of plant species dispersed by humans in the flora of Halle increased over time, but also animal-dispersed species seem to have profited from urbanization. As discussed in Chapter I, domestic animals like cats and dogs are very frequent in urbanized areas and potential dispersers of plant seeds. Species with adhesive dispersal might not only be dispersed by animals but also by humans and vehicles (Hodkinson & Thompson 1997; von der Lippe & Kowarik 2007; 2008). Additionally, species with fleshy fruits, being dispersed after digestion, are as well attractive to gardeners, if fruits are colored and showy. There are many birds in cities that disperse fleshy fruits. The decrease of wind-dispersed species might be due to unsuitable conditions in urban areas, where calms are more frequent than in the countryside (Kuttler 1993) and seeds do not reach the lee of walls and houses (see Chapters I and II). However, Lososová *et al.* (2006) found wind-dispersed species to be more frequent in urban than in agricultural habitats and assigned this pattern to the fragmentation and dynamic nature of urban landscapes. Therefore, it is likely that wind-dispersed species were introduced less often to Halle than expected because they are less attractive for planting than animal-dispersed species with fruits or cones.

The decrease of agricultural land use and the increase of urban land use in the study area are also likely to have increased the proportion of plants with hygromorphic leaves being sensitive to drought, which is in accordance with the results of Chapter II: Arable fields in the study area mostly were not irrigated, but gardens, parks, and cemeteries often are; these kind of habitats are also more shady than open arable land. The decrease of plants with hydromorphic leaves might be a result of industrial contamination of water bodies.

Surprisingly, species that grow preferably on inundated soils and aquatic plants increased their relative proportion in the flora of Halle, although the Saale river, a main habitat for aquatic plants in the city, was biologically dead in the 1970ies and 1980ies (Stolle & Klotz 2004). On the one hand, this increase is due to introduced neophytes such as *Azolla filiculoides* Lam. or *Elodea canadensis* Michx.; but also species native to Germany introduced or immigrated after the first time period (e.g. *Eleocharis palustris* L.). On the other hand, only one of the respective species present in 1687-1689 got extinct: *Equisetum fluviatile* L. was not recorded for the last time period of 2000-2008. Consequently, the species survived the heavy pollutions of the 1970ies-80ies in other aquatic habitats than the rivers: For example, several open pits from lignite mining in the city area were flooded after the end of usage and increased the available habitat for aquatic species. Earlier, the drainage of bogs for coal mining probably caused the extinction of many plants that grow in bogs (e.g. *Drosera rotundifolia* L., *Eriophorum angustifolium* Honck. or *Orchis palustris* Jacq.).

The proportions of life span types and categories of specific leaf area were stable between 1687 and 2008. Therefore, being annual, biennial or pluriennial or having a certain SLA seems to be less susceptible to changes in land use or other environmental changes. However, it might as well be that these traits are similarly influenced by urban and agricultural land use, which both put strong disturbances on vegetation: In Chapter I, SLA only differed between urbanized and semi-natural grid-cells, not between urbanized and agricultural grid-cells. Land use in the Halle region mainly changed from agricultural to urban, and agriculture might have affected the SLA of species long before urbanization did.

Differences in the analyses of associations with time and analyses of trends in trait state ratios result from the fact that associations were tested with absolute species numbers while trends were tested with ratios of relative species numbers. There might be gains in species numbers of a certain trait state but at the same time a decrease of the proportion of this trait state in the total flora because another state of the same trait gains even more species and thus increases its proportion. Therefore, proportions give a clearer picture of how urbanization affects the functional composition of the flora than absolute species numbers.

Some of the changes in the functional composition of the flora of Halle that took place in the last 320 years can also be observed in space when comparing recent urban and rural floras: Human-dispersed species, animal dispersed species, and therophytes for example, are more frequent in urban than in rural areas in Germany (Chapters I, II). However, insect-pollinated plants and species with hygromorphic leaves seem to be less frequent in urban than in rural areas (Lososová et al. 2006; Chapter I), in contrast to the increase of these species in Halle over time. Differences in scale might yield different results – Lososová et al. (2006) worked with plots ≤ 100 m²; in Chapter I we investigated grid-cells sized c. 130 km². Studies working at the scale of single cities and their surroundings found patterns similar to our temporal patterns: Wittig and Durwen (1982) for example found more species preferring nitrogen-rich, dry, warm habitats in cities in the West of Germany. The relation between species richness and human presence changes with scale (Pautasso 2007); the same could be true for relations of trait state frequency and urban land use as Chapter II suggested, but this needs further testing.

Our study shows the vast changes in an urban flora caused by the influence humans had and still have on biodiversity by intentionally (gardeners preferences; cf. Niinemets & Peñuelas 2008) or unintentionally (changing environmental conditions) selecting specific functional plant types and thus changing the composition of the flora. Despite uncertainties, historical floras provide unique opportunities to analyze past changes in biodiversity and to show trends that might proceed in the future.

Chapter V – Challenging Urban Species Diversity: Contrasting Phylogenetic Patterns across Plant Functional Groups in Germany

1. Introduction

The high vascular plant species richness of urbanized areas in Germany is biased towards species with specific functional traits or trait states, and towards common species, as shown in the preceding chapters. The last aspect of species diversity to be analyzed here is phylogenetic diversity: In terms of species richness, an assemblage of three *Poaceae* species seems as diverse as an assemblage of one *Poaceae*, one *Asteraceae* and one *Fagaceae* species; but the former assemblage appears much less diverse when considering their phylogenetic background: The three *Poaceae* species belong to one family and are thus closer related to each other than the species from the three families of *Poaceae*, *Asteraceae* and *Fagaceae*. Phylogenetic diversity, which measures the diversity of evolutionary relationships between species, reveals these underlying patterns, and so provides valuable information for species conservation and about mechanisms of species assembly (Vane-Wright *et al.* 1991).

Phylogenetically closely related species often share specific traits or trait states through their common origin and evolutionary history (evolutionary niche conservatism; Harvey & Pagel 1991; Prinzing *et al.* 2001).

Hence, phylogenetic diversity is usually interrelated with the frequency of species per functional trait. However, phylogenetically closely related species can also develop different trait states due to adaptive radiation (e.g. Schluter 2000; Ackerly & Nyffeler 2004; Prinzing *et al.* 2008). In both cases, the environment influences the functional and phylogenetic structure of a species assemblage. We therefore expect differences not only in the functional but also in the phylogenetic structure of floras from urbanized and non-urbanized areas. While influences of urbanization on functional traits have been confirmed for a range of plant traits (e.g. Kleyer 2002; Williams *et al.* 2005; Lososová *et al.* 2006; present study), little is known about the effects of urbanization on phylogenetic diversity (but see Ricotta *et al.* 2008a).

Here, we compare the phylogenetic diversity of German vascular plant as-semblages between urbanized and two types of non-urbanized areas, i.e. agricul-tural and semi-natural including forests (both referred to as rural, as defined in Chapter I, see Fig. A2). Our approach is a macroecological one, suitable to re-veal large-scale patterns and well suited to reflect the influence of urbanization on biodiversity that does not stop at city borders but acts on large areas. More-over, the positive relation between urban land use and species richness is espe-cially strong at coarse scales (Pautasso 2007). We used the gridded dataset from Chapter I, for which previous analyses have shown that the species richness of vascular plants is higher in urbanized than in rural grid-cells (Kühn *et al.* 2004a). Phylogenetic diversity in areas with dense human population might be even higher than expected from species numbers, as Sechrest *et al.* (2002) have shown for carnivores and primates in areas that are naturally species rich. Corre-spondingly, urbanized areas could be expected to have a higher phylogenetic diversity than rural areas, because heterogeneous landscapes provide a variety of niches for a variety of lineages (cf. Ricotta *et al.* 2008a), while agricultural land-scapes are homogeneous over large areas. On the other hand, if a trait is highly conserved, then the urban environment should filter for closely related species: Species groups characterized by a conservative trait that is suitable for urban environments should be phylogenetically clustered within urbanized landscapes (e.g. Cavender-Bares *et al.* 2006; Swenson *et al.* 2007).

There are environmental filters (Zobel 1997) in both urbanized and rural areas that might restrict species richness to species capable of passing the filters, i.e. plants with suitable trait states, as shown in the preceding chapters. These filters, such as the fragmentation of urbanized landscapes or the regular disturbance in agricultural landscapes, might increase the phylogenetic diversity of plants with well suited traits but decrease the phylogenetic diversity of plants with less suited traits. Plants well suited for urban environments should be able to colonize a range of urban habitats (cf. Kowarik 2008) and thus to establish a high phylogenetic diversity. Plants less suited for urban environments should be restricted to only a few urban habitats and consequently have a restricted phylogenetic diversity.

We therefore tested the hypothesis that species-rich urbanized areas, given their high geological and structural heterogeneity, also have a higher phylogenetic diversity than rural areas. This should apply because the higher urban habitat heterogeneity is expected to hold a higher number of different lineages. We further tested whether species richness and phylogenetic diversity patterns differ systematically between species groups characterized by different trait states, i.e. whether species richness or phylogenetic diversity are higher in groups with traits suitable for urban environments and lower in groups with more unsuitable traits.

2. Materials and Methods

2.1. Data Sources

Plant species occurrences were calculated at the same *c.* 12 km × 11 km grid from FLORKART used in Chapters I and III, divided into urbanized, agricultural and semi-natural grid-cells as in Chapter I (Fig. A2). We again only referred to occurrences of the spontaneous flora in sufficiently mapped grid-cells (see Chapter I, Materials and Methods).

Data on species traits and phylogeny originate from BiolFlor (Klotz *et al.* 2002; Kühn *et al.* 2004b; http://www.ufz.de/biolflor); the trait 'dispersal type'

was taken from LEDA (Kleyer *et al.* 2008; http://www.leda-traitbase.org/
LEDAportal). We used traits for which the analyses in Chapter I yielded distinct
urban-rural patterns (Table A1).

The same environmental parameters on climate, topography, soils, and geology used in Chapter I (Table A2) were used in addition to the three land-use
types (agricultural semi-natural, urban) to explain phylogenetic diversity.

2.2. Data Analyses

We first calculated the species richness of the total flora and of each species
group with a specific functional trait. We tested whether the groups reflect the
richness pattern of the total flora or vary according to their trait state. To control
for spatial autocorrelation among grid-cells and for effects of environmental
parameters other than land use (Table A2), we developed two spatial autoregressive error models (Dormann *et al.* 2007; Kissling & Carl 2008). One was an
intercept-only model with species richness as the response (SARerr-null; correcting only for spatial autocorrelation), the other had species richness as response
and climate, topography, soil, and geology as explanatory variables (SARerr-
env; correcting for both spatial autocorrelation and environmental parameters).
The lag-distance for which we considered the influence of autocorrelation was
2.5 grid-cells. We did not include land-use types in the models because of a
highly imbalanced sampling design (1365 agricultural, 312 semi-natural, 59
urbanized grid-cells). We instead calculated the residuals from SARerr-null and
SARerr-env models and assessed the effect of land-use type on the species richness of the total German flora and the 25 species groups by a resampling approach. We calculated the mean of the models' residuals per grid-cell type (agricultural, semi-natural, and urbanized) and separately resampled, according to the
number of urbanized grid-cells, 59 randomly chosen agricultural or semi-natural
grid-cells 999 times. We then tested for significant differences in the residuals'
mean values between urbanized and agricultural grid-cells and between urbanized and semi-natural grid-cells using the z-statistic (comparison of one value to
a distribution of values; *P*-values calculated for absolute z standard normal devi-

ates). This is exactly the procedure used in Chapter I but with spatial autoregressive error models instead of multiple linear models.

To compare phylogenetic diversity among the land-use types, we combined the matrices on species per grid-cell and on phylogenetic code per species to calculate phylogenetic diversity per grid-cell. The phylogenetic code of a species (as assigned to each species in BiolFlor; Durka 2002; Kühn *et al.* 2004b) marks its position in the phylogenetic tree and therefore its position relative to other species in the tree. In BiolFlor, only the topology of a species tree is given, not the branch lengths. Consequently, all branches are treated as having the same length and the phylogenetic distance between species can be derived from the number of nodes separating one species from another. This is a good alternative for the calculation of phylogenetic diversity indices if exact branch lengths are unknown (Faith 1992). Since our aim was to disentangle the effects of species richness and phylogenetic diversity, we used average taxonomic distinctness (Δ^+) following Warwick and Clarke (1982). Δ^+ is unbiased by species richness, i.e. it does not automatically increase with increasing species richness. There are several mathematically related indices such as Rao's quadratic entropy (Rao 1982) or Webb's Net Relatedness Index (NRI; Webb *et al.* 2002). However, NRI was defined for slightly different questions, as it quantifies the distribution of taxa in a sample relative to a pool. Additionally, a comparative study shows that only Warwick and Clarke's Δ^+ is exactly independent of species richness and reflects the phylogenetic structure of a subset from a phylogenetic tree best (Schweiger *et al.* 2008). Δ^+ originally was developed on taxonomic relationships but it can be easily adapted to phylogenetic information by substituting the taxonomically weighted distance by phylogenetic distance (see also Schweiger *et al.* 2008). The index was calculated as

$$\Delta^+ = [\Sigma\Sigma_{i<j}\, d_{i,j}] / [s(s-1)/2]$$

where $d_{i,j}$ is the distance matrix of nodes and s is the number of species. Thus, the index is based on a pairwise distance matrix defined by the number of nodes that separate one species from another and can be interpreted as the mean distance between two randomly chosen species independent from their distance from the root of the tree.

We calculated Δ^+ per grid-cell; first for the total German flora, then for 25 groups of species characterized by a specific trait state, e.g. for all insect-, self- or wind-pollinated species or for all species with scleromorphic or hygromorphic leaves (see Table A1). For all species groups, both native and exotic species were considered; except, of course, when grouping was based on natives, archaeophytes and neophytes. As for species richness, SARerr-null models corrected for effects of spatial autocorrelation on Δ^+ and SARerr-env models corrected for effects of spatial autocorrelation, climate, topography, soil, and geology. We also resampled the models' residuals for agricultural and semi-natural grid-cells and compared the residuals' mean values for urbanized, agricultural and semi-natural grid-cells with the z-statistic.

If SARerr-null and SARerr-env models yield the same result, e.g. higher urban than rural phylogenetic diversity, then the parameters causing this pattern should be 'urban-intrinsic' (or 'rural-intrinsic'). Examples for urban-intrinsic parameters are the density of built-up area or disturbance intensity. If the SARerr-null model shows a difference between urbanized and rural areas but the SARerr-env model does not, then the differences shown by the former could be explained by the parameters accounted for in the latter, such as higher temperatures in urbanized areas (Fig. A1). If the SARerr-null model shows no differences between urbanized and rural areas but the SARerr-env model does, then 'urban-intrinsic' and environmental parameters are operating in opposite directions.

All analyses in this chapter were done with version 2.6.0 of R (http://www.R-project.org; R Development Core Team 2007).

3. Results

3.1. Species Richness and Phylogenetic Diversity

Species richness was significantly higher in urbanized than in agricultural or semi-natural grid-cells, not only regarding the total flora but also throughout all tested trait state groups (Table 5.1). This was true when only accounting for spatial autocorrelation as well as when accounting for both spatial autocorrelation and environmental variables. Despite this high urban species richness, the phylogenetic distinctness of the total flora was not higher in urbanized than in rural areas (Fig. 5.1a; Table 5.1) but rather showed a tendency towards being decreased. The phylogenetic diversity of the total flora was higher in semi-natural than in urbanized grid-cells in the SARerr-null model but showed no differences between the urbanized and the two types of rural grid-cells in the SARerr-env model.

3.2. Phylogenetic Diversity across Plant Functional Groups

For particular species groups, different patterns occurred according to their trait states. Phylogenetic distinctness of hydrochorous species, geophytes, hydrophytes, phanerophytes, and plants with hygromorphic, mesomorphic, summer green or spring green leaves was highest in semi-natural grid-cells in both the SARerr-null and the SARerr-env models (Fig. 5.1b-i, Table 5.1; but see hydrophytes for urbanized and agricultural grid-cells in the SARerr-env models). The phylogenetic distinctness of human-dispersed species and pluriennial plants was highest in semi-natural grid-cells in the SARerr-null models, and highest in agricultural grid-cells for helomorphic plants in the SARerr-env models (Table 5.1). Plants that have a biennial life cycle, scleromorphic, succulent, evergreen or overwintering green leaves or species that are neophytes or self- or wind-pollinated had a higher phylogenetic distinctness in urbanized than in agricultural and semi-natural grid-cells in both the SARerr-null and SARerr-env models (Fig. 5.1j-q, Table 5.1; but see overwintering green leaves for urbanized and semi-natural grid-cells in the SARerr-env models). Animal-dispersed species had a slightly higher phylogenetic diversity in urbanized than in agricultural grid-cells in the SARerr-env models (Fig. 5.1r, Table 5.1).

4. Discussion

Our study highlights a pronounced discrepancy between species richness and phylogenetic diversity in urbanized areas. Generally, we expect phylogenetic diversity to be higher in heterogeneous than in homogeneous landscapes, because the former provide more niches for a variety of lineages (Ricotta *et al.* 2008a). Due to the high heterogeneity of urbanized landscapes (Niemelä 1999) and because modern agricultural habitats as well as heavily managed forested habitats are very homogeneous, we expected phylogenetic diversity to be higher in urbanized than in rural areas. However, our results suggest the opposite: Phylogenetic diversity does not reflect the high species richness of urbanized areas.

Moreover, when changing perspective from the total flora to species groups, it is apparent that the patterns of phylogenetic diversity differ between species groups characterized by specific functional traits. Using different approaches, studies have shown that the phylogenetic structure of a community can depend on taxonomic or spatial scale (Cavender-Bares *et al.* 2006 for all seed plants and single lineages in Floridian plant communities; and Swenson *et al.* 2007 for size classes of tropical trees). Our results suggest the presence of selective environmental filters in urbanized areas which differ from those of rural areas. When species are assembled from the species pool, they have to pass a series of filters whose properties determine the structure of the assemblage according to species-specific trait compositions (e.g. Zobel 1997; Schweiger *et al.* 2005). The urban filters act on all species but depending on their trait states, some plants are able to pass the filters, while others are not. The reduction of phylogenetic diversity in urbanized areas may then be caused by the presence of only particular species groups that can be regarded as adapted to non-urban conditions (species groups in Fig. 5.1b-i).

Table 5. 1 – Differences in mean between the species richness and phylogenetic diversity of the flora of urbanized, agricultural, and semi-natural grid-cells in Germany
See below for further explanation.

Trait	Trait state	Species richness – SARerr-null		Species richness – SARerr-env		Phylogenetic diversity – SARerr-null		Phylogenetic diversity – SARerr-env	
	Total flora	a < u***	sn < u***	a < u***	sn < u***	a = u	sn > u+	a = u	sn = u
Dispersal type	Anemochory	a < u***	sn < u***	a < u***	sn < u***	a = u	sn = u	a = u	sn = u
	Hemerochory	a < u***	sn < u***	a < u***	sn < u***	a > u+	sn > u*	a = u	sn = u
	Hydrochory	a < u***	sn < u***	a < u***	sn < u***	a > u*	sn > u***	a > u***	sn > u***
	Zoochory	a < u***	sn < u***	a < u***	sn < u***	a > u*	sn = u	a < u+	sn = u
Floristic status	Archaeophyte	a < u***	sn < u***	a < u***	sn < u***	a = u	sn = u	a = u	sn = u
	Native	a < u***	sn < u***	a < u***	sn < u***	a = u	sn = u	a = u	sn = u
	Neophyte	a < u***	sn < u***	a < u***	sn < u***	a < u***	sn < u***	a < u**	sn < u***
Leaf anatomy	Helomorphic	a < u***	sn < u***	a < u***	sn < u***	a = u	sn = u	a > u*	sn > u+
	Hydromorphic	a < u***	sn < u***	a < u***	sn < u***	a = u	sn < u+	a = u	sn = u
	Hygromorphic	a < u***	sn < u***	a < u***	sn < u***	a > u*	sn > u***	a = u	sn > u**
	Mesomorphic	a < u***	sn < u***	a < u***	sn < u***	a > u*	sn > u**	a > u+	sn > u**
	Scleromorphic	a < u***	sn < u***	a < u***	sn < u***	a < u**	sn < u***	a < u*	sn < u**
	Succulent	a < u***	sn < u***	a < u***	sn < u***	a < u**	sn < u***	a < u**	sn < u***
Leaf persistence	Evergreen	a < u***	sn < u***	a < u***	sn < u***	a < u+	sn < u***	a < u**	sn < u***
	Spring green	a < u***	sn < u***	a < u***	sn < u***	a > u***	sn > u***	a > u***	sn > u***

Trait	Trait state	Species richness – SARerr-null		Species richness – SARerr-env		Phylogenetic diversity – SARerr-null		Phylogenetic diversity – SARerr-env	
Leaf persistence	Summer green	a < u***	sn < u***	a < u***	sn < u***	a > u***	sn > u***	a > u +	sn > u*
	Overwintering green	a < u***	sn < u***	a < u***	sn < u***	a < u***	sn < u**	a < u**	sn = u
Life form	Chamaephytes	a < u***	sn < u***	a < u***	sn < u***	a = u	sn < u**	a = u	sn = u
	Geophytes	a < u***	sn < u***	a < u***	sn < u***	a > u*	sn > u***	a = u	sn > u*
	Hemicryptophytes	a < u***	sn < u***	a < u***	sn < u***	a < u*	sn = u	a < u**	sn = u
	Hydrophytes	a < u***	sn < u***	a < u***	sn < u***	a > u*	sn > u***	a = u	sn > u*
	Phanerophytes	a < u***	sn < u***	a < u***	sn < u***	a = u	sn > u***	a = u	sn > u**
	Therophytes	a < u***	sn < u***	a < u***	sn < u***	a = u	sn = u	a = u	sn = u
Life span	Annual	a < u***	sn < u***	a < u***	sn < u***	a = u	sn = u	a = u	sn = u
	Biennial	a < u***	sn < u***	a < u***	sn < u***	a < u***	sn < u**	a < u***	sn < u***
	Pluriennial	a < u***	sn < u***	a < u***	sn < u***	a > u*	sn > u***	a = u	sn = u
Pollen vector	Insects	a < u***	sn < u***	a < u***	sn < u***	a = u	sn < u*	a = u	sn = u
	Selfing	a < u***	sn < u***	a < u***	sn < u***	a < u*	sn < u***	a < u**	sn < u***
	Wind	a < u***	sn < u***	a < u***	sn < u***	a < u***	sn < u***	a < u***	sn < u***

Further explanation for table 5.1.: u = urbanized; a = agricultural; sn = semi-natural. SARerr-null is corrected for spatial autocorrelation; SARerr-env is corrected for spatial autocorrelation and environmental variables (see *Materials and Methods* section of this chapter for details). *P*-values: $0.05 < P \leq 0.1+$, $P \leq 0.05*$, $P \leq 0.01**$, $P \leq 0.001***$, for non significant differences, equal values are assumed.

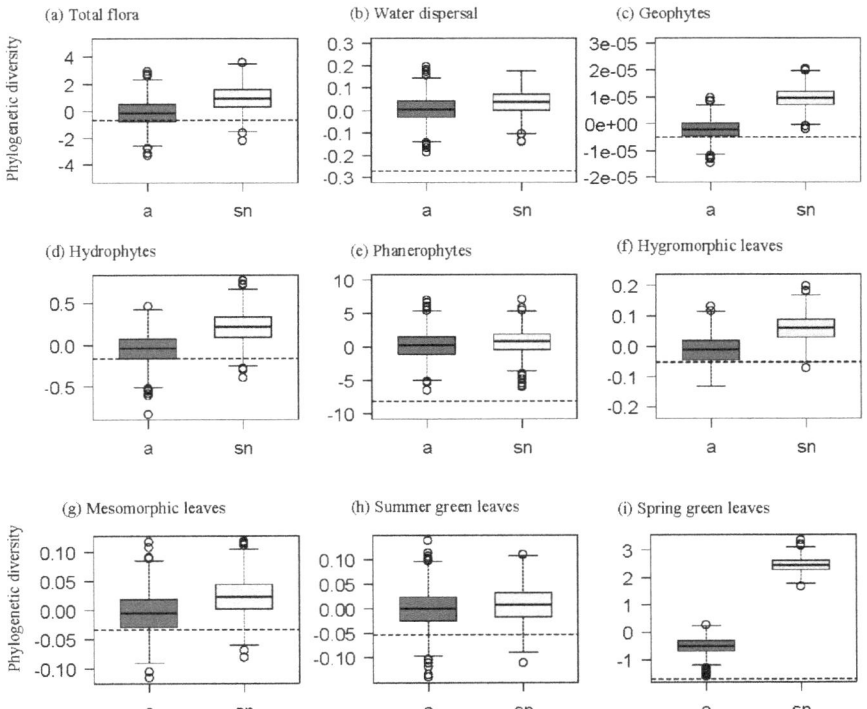

Figure 5. 1 – Phylogenetic diversity of the flora in urbanized, agricultural and semi-natural grid-cells in Germany

Selected results are shown for the total flora and for species groups that share the same type of a functional trait: (a) all plant species, (b) – (i) species groups that share the pattern or tendency of the total flora: (b) hydrochorous species; (c) geophytes; (d) hydrophytes; (e) phanerophytes; (f) species with hygromorphic leaves; (g) species with mesomorphic leaves; (h) species with summer green leaves; (i) species with spring green leaves. (j) – (r) species groups with a higher urban than agricultural or semi-natural phylogenetic diversity: (j) biennials; (k) species with scleromorphic leaves; (l) species with succulent leaves; (m) species with evergreen leaves; (n) species with overwintering green leaves; (o) neophytes; (p) self-pollinated species; (q) wind-pollinated species; (r) species dispersed by animals. Boxplots represent median (line), 25-75 % quartiles (boxes), ranges (whiskers) and extreme values (circles). Dark grey = agricultural grid-cells; light grey = semi-natural grid-cells; dashed line = urbanized grid-cells. Values for agricultural and semi-natural grid-cells are based on resampling. Shown are residuals (see *Materials and Methods* section of this chapter for details). *P*-values for differences between urbanized and agricultural/urbanized and semi-natural grid-cells are shown in Table 5.1.

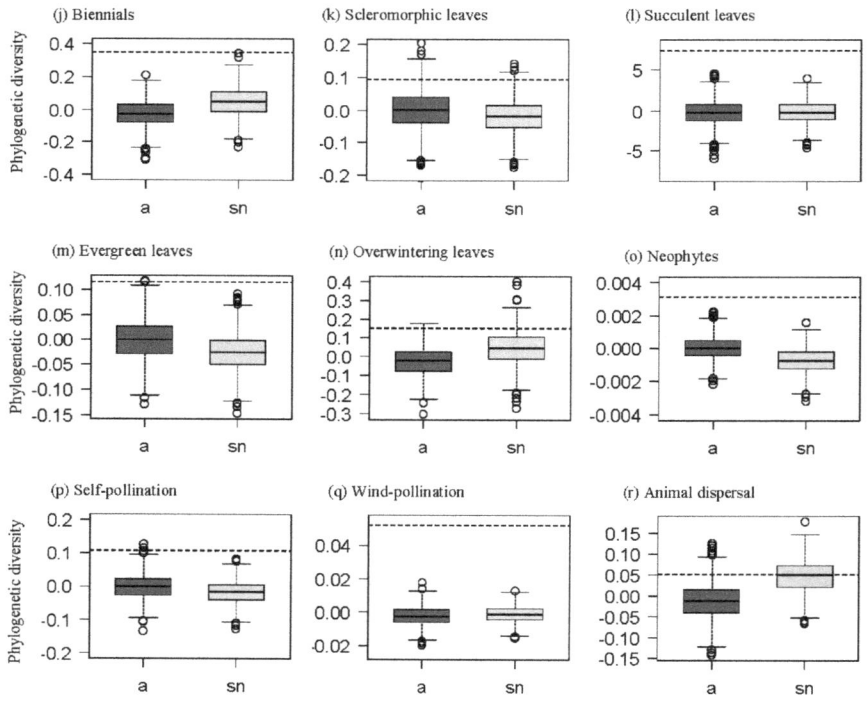

Figure 5. 1. – continued

Some of these groups have not only the highest phylogenetic diversity in semi-natural areas but also the highest frequency in the German flora, e.g. pluri-ennials or plants with mesomorphic leaves (Fig. 5.2). This partly explains why

they reflect the pattern of the total flora (or vice-versa). Furthermore, some species groups are intercorrelated, e.g. phanerophytes are usually pluriennial. Therefore, the low phylogenetic diversity of pluriennials in urbanized grid-cells might
mainly reflect the even more significant reduction in urbanized phylogenetic
diversity for phanerophytes.

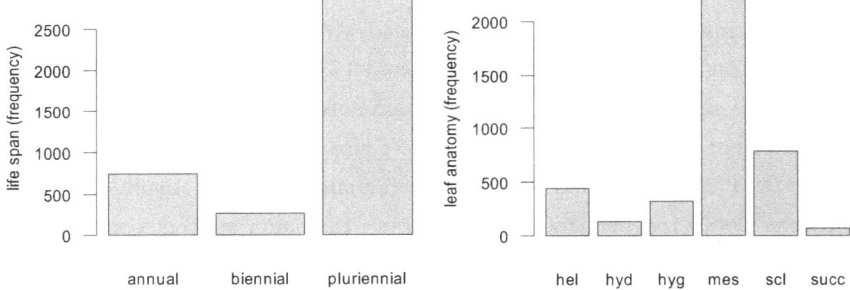

Figure 5. 2 – Frequency of trait states in the German flora for the traits life span and leaf
anatomy
Left: life span; Right: leaf anatomy with hel = helomorphic; hyd = hydromorphic; hyg =
hygromorphic; mes = mesomorphic; scl = scleromorphic; succ = succulent

    Species groups with a reduced phylogenetic diversity in rural grid-cells (Fig.
5.1j-r) might be more sensitive to non-urban filters, e.g. higher competition due
to low disturbance frequency. However, they are adapted to urban conditions,
such as biennial or wind-pollinated plants which are adapted to disturbance and
fragmentation (Lososová *et al.* 2006). These species consequently are more
frequent in urbanized than in rural areas. But why are they also phylogenetically
more diverse in urbanized grid-cells?

    The phylogenetic urban-rural patterns may result from a combination of phe-
notypic clustering and phylogenetic overdispersion (Cavender-Bares *et al.*
2006). Plants that share trait states are able to pass the same environmental filters
and thus tend to occur in similar habitats. For example, biennials occur more
often in cities and pluriennials more often in semi-natural habitats; they are phe-
notypically clustered. However, if species are too similar because they share

critical trait states, they cannot coexist (e.g. Chesson 2000; Prinzing *et al.* 2008). Accordingly, the phylogenetic overdispersion within species groups character-ized by traits suitable to pass the urban (or rural) environmental filters can be due to several mechanisms: Firstly, the species within a species group do not share trait states because of trait conservatism within lineages (e.g. Prinzing *et al.* 2001) but due to convergent trait evolution (Cavender-Bares *et al.* 2006). Secondly, closely related species get replaced with less related species, i.e. com-petitive exclusion limits similarity (Diamond 1975, cit. in Helmus *et al.* 2007; Swenson *et al.* 2007). However, competition acts on a much smaller scale than the one of our study and is thus unlikely to cause the patterns we found. Thirdly, similar and closely related species differentiated through adaptive radiation (Schluter 2000) independent of urbanization. Radiation enables species to use different resources within the same environment, e.g. in an urbanized landscape. This means that plants with traits well-suited for urbanized areas find several niches and thus can contribute to a high phylogenetic diversity of the urban flora. Conversely, plants with traits less well adapted to the urban environment can only grow in a few urban habitats and thus only contribute with a reduced phylogenetic diversity to the urban flora. With this last point, we can explain best why species richness is high in urbanized grid-cells, while phylogenetic diversity is reduced within the total flora but increased within 'urban-adapted' species groups. The high urban species richness seems to rest upon single (or few) but speciose lineages characterized by conservative traits that enable them to pass the urban filters and to settle many different habitats within a city. Grasses for example are very speciose and wind-pollinated (Gorelick 2001), a trait that is phylogenetically conserved (Chazdon *et al.* 2003). Similarly, the genus *Oenothera* is a speciose lineage especially occurring in urban habitats (cf. Sukopp *et al.* 1979). Most *Oenothera* species are biennials and hemicrypto-phytes, which are both traits with a higher phylogenetic diversity in urbanized grid-cells (Table 5.1). Biennials and hemicryptophytes, in turn, are usually non-woody, a trait that is also phylogenetically conserved (Ackerly & Donoghue 1995). Cavender-Bares *et al.* (2006) and Swenson *et al.* (2006) showed that phylogenetic overdispersion is more likely at finer taxonomic levels. In our study, phylogenetic overdispersion occurs at the level of species groups characterized by certain trait states, which is finer than the level of the total flora.

Differences between models correcting for spatial autocorrelation (SARerr-null) and models correcting for both spatial autocorrelation and environmental variables (SARerr-env) indicate whether the environmental variables that we accounted for or 'urban-intrinsic' filters influence phylogenetic diversity. SARerr-null and SARerr-env models yielded no fundamental differences across most species groups (Table 5.1). Therefore, the acting filters seem to be 'urban-intrinsic' (i.e. they were not accounted for in the model, e.g. strong fragmentation of habitats by high proportions of built-up area, high disturbance frequencies, and high temporal land-use turnover; Sukopp *et al.* 1979; Kleyer 2002). The phylogenetic diversity of the total flora, human-dispersed species, insect-pollinated plants, chamaephytes, geophytes, and plants with hydro- or hygromorphic leaves differs between urbanized and agricultural or urbanized and semi-natural grid-cells in the SARerr-null models. One or both of these differences disappear in the SARerr-env models (Table 5.1). Thus, the climatic, topographic, edaphic and geologic variables that we accounted for are relevant for differences in the phylogenetic diversity of these species groups between urbanized and non-urbanized grid-cells. For example, increased temperatures and reduced rainfall in urbanized grid-cells (Fig. A1) are probably major filters reducing the phylogenetic diversity of hygromorphic plants in urbanized areas: They restrict these plants to special habitats within a city, such as urban parks along rivers which are cooler and more humid and shady than other urban habitats. Regarding hydrochorous species, self-pollinated plants, plants with helomorphic, succulent or evergreen leaves, biennials, and hemicryptophytes, it seems that the environmental filters and the 'urban-intrinsic' or 'rural-intrinsic' filters partly mask each other (Table 5.1): Differences in phylogenetic diversity between urbanized and rural grid-cells are even stronger in the SARerr-env than in the SARerr-null models. Most of these species groups have the highest phylogenetic diversity in urbanized grid-cells. The SARerr-null and SARerr-env models differ with respect to differences between urbanized and agricultural grid-cells (except for helomorphic plants and biennials; Table 5.1). Succulent leaves for example, might be especially suitable with respect to the relatively warm and dry city climate (Fig. A1) and therefore develop a high phylogenetic diversity in urbanized grid-cells. They might as well be suitable with respect to the high density of sealed surfaces in urbanized areas and the accordingly high proportion of surface runoff that increases the aridity of urban habitats. Surface runoff would then be an 'urban-intrinsic' filter not accounted for in the SARerr-env models. Its effects might be masked by the effects of high temperatures and reduced precipitation.

With grid-cells sized c. 12 km × 11 km, the scale of the study is fairly large. As biodiversity patterns can vary between large and small spatial scales (e.g. Cavender-Bares et al. 2006; Pautasso 2007), the results of our analyses might change when tested on a smaller scale. Unfortunately, plant occurrence data for the whole area of Germany are not available for a higher resolution than the grid used. Further research is needed to clarify whether the patterns we found are robust over varying scales.

Our study shows that the generally high plant species richness of urbanized areas is not reflected in phylogenetic diversity but is mainly due to more closely related plants with (pre-) adaptations to urban environments. The loss of phylogenetic information decreases the capacity of species assemblages to respond to environmental changes and might negatively affect ecosystem functioning (e.g. Maherali & Klironomos 2007). Therefore, nature conservation should, besides the number and identity of species, also account for phylogenetic diversity to sustain the capacity of species assemblages to respond to changing environmental conditions. As urbanization is unlikely to stop, we need strategies for protecting biodiversity in spite of urbanization, i.e. also within urbanized areas. To give valuable recommendations for the protection of biodiversity in urbanized areas, we need further analyses that explore the phylogenetic diversity of semi-natural vs. typical urban habitats within urbanized areas. Such studies might assess the potential of semi-natural areas within urbanized landscapes in conserving a high phylogenetic diversity (across all functional groups). The consideration of both the phylogenetic relationships and traits of species in addition to species richness is crucial for a detailed understanding of how species assemblages develop and change with a changing environment.

Synthesis and Conclusions

1. Synthesis

1.1. Urban Plant Biodiversity Research up to now

The environmental conditions of towns and cities clearly differ from those in rural areas: Climate (Landsberg 1981; Oke 1982; Kuttler 2008), soils (Sukopp *et al.* 1979; Blume 1998), hydrological conditions (Wessolek & Renger 1998; Paul & Meyer 2001), disturbance regimes and land-use structure (Niemelä 1999) change along urbanization gradients. The characteristics of the urban environment have been the focus of early ecological studies aiming at an inventory and characterization of urban ecosystems. The remarkable differences between urban and non-urban ecosystems should result in a species turnover along urbanization gradients. In fact, there are species that only occur outside of cities (urban avoiders; see Blair 1996; urbanophobic species; see Wittig 2002) and species that preferably occur within cities (urban exploiters; see Blair 1996; urbanophilic species; see Wittig 2002). Moreover, there are even plant families that hardly occur in urbanized areas, e.g. *Cyparaceae* or *Orchidaceae*, while other families cumulate there, e.g. *Onagraceae* or *Polygonaceae* (Wittig 2002). This taxonomic turnover should be accompanied by a turnover of species traits and ecological preferences according to the ecological changes along urbanization gradients. Such a turnover has been proven for several indicator-value spectra of vascular plants: Species indicating e.g. high temperatures, low moisture or high nitrogen contents of soils increase in proportion in urbanized areas (Wittig & Durwen 1982; Klotz 1989; Chocholoušková & Pyšek 2003). Also plant species adapted to disturbance, such as short-lived species preferably occur in cities (Wittig 2002). However, annual species often are even more frequent in agricultural than in urban habitats (Lososová *et al.* 2006).

Although several studies have shown such changes in the composition of floras with increasing urbanization, most of them concentrated on indicator-values
and less dealt with biological traits, such as type of pollination, dispersal or reproduction. For some biological traits, inventories were made for selected species (e.g. Wittig 2002 for 20 extremely urbanophilic species), but only few statistical analyses were performed for the trait state frequency of a total urban vs.
rural flora (but see Lososová et al. 2006 for urban and agricultural vegetation
plots in the Czech Republic). Most studies exemplarily investigated single towns
or cities (e.g. Wittig & Durwen 1982 for Bielefeld, Dortmund, Köln and Müns-
ter; Godefroid 2001 for Brussels; or Kleyer 2002 for Stuttgart). Temporal studies
on changes of urban plant assemblages hardly ever exceeded the 19th century
(e.g. Chocholoušková & Pyšek 2003: 1880-2000; Van der Veken et al. 2004:
1880-1999; Tait et al. 2005: 1836-2002; Lavergne et al. 2006: 1886-2001).
Moreover, due to the lack of phylogenies for many taxa, phylogeny seldom has
been used for comparative studies; mostly taxonomies have been used (e.g.
Crawley et al. 1996 by comparing the representation of plant families among
native and alien species of the British Isles; Pyšek 1998 by investigating the
distribution of alien plant species within families and higher taxonomic units;
Ricotta et al. 2008a by comparing local and regional plant species pools in the
city of Rome). Although taxonomies approximate phylogenies, the latter give
more detailed information: Taxonomies are based on the Linnean system that
ranks species into genera, genera into families, etc. Therefore, taxa of the same
hierarchical level count equally even if the higher taxa to which they belong
differ significantly in species richness (cf. Ricotta et al. 2008a). Contrastingly,
phylogenies show the relatedness of species without grouping them into higher
taxa.

1.2. Contributions of this Study to Urban Plant Biodiversity Research

Our study for the first time investigated effects of urbanization on a large
number of plant traits throughout Germany, using extensive databases that cover
most of the German flora. These databases, especially BiolFlor and LEDA, be-

came available only a few years ago – the BiolFlor database was released in 2002 (see Kühn *et al.* 2004b); the LEDA-project ran from 2002 to 2005. Now, with these extensive databases available, it was a logical consequence to test patterns known from smaller scales, single cities or selected groups of species for all of Germany and to expand analyses to further traits to show which plant traits are associated with successful performance in urbanized and urbanizing areas. Similar to traits, the phylogenetic code from BiolFlor, which is available for most species of the German flora (more than 3000 species; Durka 2002), made it possible to go beyond taxonomy and to analyze the phylogenetic structure of the total German flora and of several trait-groups. Also the extensive historical records on plant occurrences in Halle have no equivalent in Central Europe. They allowed for the investigation of the effect of 320 years of urbanization on an urban flora.

It has been known since the 1970ies that urbanized areas are richer in species than their rural surroundings (Walters 1970; Haeupler 1975), but Kühn *et al.* (2004a) were the first to prove it for all of Germany, for the same resolution of grid-cells as used in this study. The analysis of Kühn *et al.* was a good starting point to further investigate whether species rich urbanized areas are biased towards species with specific traits and whether urbanized areas also harbor a high phylogenetic diversity. The present study provided these analyses and showed that urban plant species assemblages preferably consist of frequent species from specific lineages with urban-adapted trait states.

We confirmed results from smaller study areas, single cities or selected groups of species: Many plant species that grow in urbanized areas are adapted to drought and high temperatures, e.g. through leaf anatomy, short life cycles that enable the avoidance of unfavorable seasons or indirectly through the lack of investment in subterranean vegetative buds;

are adapted to disturbance, e.g. through short life cycles, reproduction by seeds or fast turnover of leaves connected with reduced investements into leaf substance; are able to grow in soils with high nitrogen contents; rely more on abiotic than on biotic pollination vectors.

Therefore, we were able to generalize these patterns from case studies to the total German flora. Furthermore, our study added information to earlier studies by showing that species deal with the fragmentation of urbanized areas by preferably using humans as dispersal vectors as well as animals that are less affected by fragmentation, such as birds.

New information was also gained from the analysis of leaf traits: Even though the gain in evergreen species in Europe due to climate warming has been broadly discussed (laurophyllization; see Klötzli & Walther 1999; Walther et al. 2007; Berger et al. 2007), we found no evidence that urbanization increases the proportion of evergreen species. Although urban areas are warmer than their surroundings and are themselves subjected to climate warming, evergreen species do not seem to profit from this. Negative effects of disturbance and stress probably outpace positive climatic effects in urbanized areas by decreasing the proportion of evergreen species: One example might be the use of chloride salts for de-icing of streets that many evergreen species are not tolerant to (Niinemets & Peñuelas 2008). The results for specific leaf area and leaf dry matter content support the results for evergreen species.

The study also added information to the fact that species in urbanized areas preferably belong to specific families by showing in more detail and on a statistical basis that the plant species of the total flora of urbanized areas are more closely related than their rural counterparts. Thus, our work supports the concept of phylogenetic niche conservatism (Prinzing et al. 2001). Simultaneously, it showed for the first time that groups of species that share trait states adapted to the characteristics of urban environments are less related to each other in urbanized areas than in non-urban areas; while groups of species that share trait states less adapted to the urban environment are closer related there than in non-urban areas. In short, we showed that groups of species that are adapted to their environment are able to support a high phylogenetic diversity there, while this is not the case for groups of species less adapted to this environment.

1.3. Notes on the Basic Framework of this Study

Mapping the total German flora is a huge task that requires lots of participants and time. Thus, restrictions have to be made, for example concerning the scale of mapping. The floristic mapping of Germany took grid-cells sized 12 × 11 km each as basis. These are the grid-cells used in this study. It is clear that analyses of such large study areas are only able to show large-scale patterns: If a species is recorded for one grid-cell, we cannot tell from the data whether it occurs everywhere in the cell or whether it is lacking in e.g. 70% of it. Moreover, the mapping provides presence-absence data but no abundance data, meaning that every species counts the same in the analyses, no matter whether it only occurs with ten individuals in a grid-cell or with thousands of individuals. Consequently, macroecological analyses like the one on the basis of grid-cells cannot replace analyses of single habitats, communities or species, which are of special interest for the conservation of species. Nevertheless, they can provide general information on the diversity and distribution of taxa and traits and on the assembly of species: We showed patterns for the total German flora that until now have only been shown for single plant associations, single cities or smaller scales. Thus, our macroecological approach made the step from case studies to a more comprehensive, far-reaching picture of urban floras.

2. Conclusions

German cities are richer in vascular plant species than their rural surroundings, but this richness is restricted in several ways: (i) Urban floras are biased towards species with specific traits that are suitable for plant life in urbanized areas; (ii) the high urban species richness is to a large extent based on common species; (iii) urban species assemblages are phylogenetically less diverse than rural species assemblages.

Most patterns presented in this book are large-scale patterns, i.e. they are no local or regional peculiarities. As the floras of cities worldwide get more and more similar to each other (McKinney 2006), patterns that are valid for Central

European cities might also hold for cities in other parts of the world, at least in temperate zones. Consequently, if urbanization has not yet caused the decline of plant species diversity worldwide, it is at least able to do so by fostering common species with "urban-suitable" traits, limiting rare species with traits less suitable for urban environments, and decreasing phylogenetic diversity.

The high species richness of urbanized areas increases regional species richness but decreases global species richness due to homogenization (Sax & Gaines 2003; McKinney 2006). Functional homogenization (Winter *et al.* 2008) and phylogenetic homogenization of the world's urban floras – and with them, of the world's total flora – might be even stronger.

Because urbanization often took place in species rich areas (Kühn *et al.* 2004a) and is still taking place in such areas (Cincotta *et al.* 2000; Balmford *et al.* 2001; Sechrest *et al.* 2002), species conservation should not only focus on natural and semi-natural landscapes but also on urbanized and urbanizing landscapes. Urbanization surely will continue (United Nations 2006), which means that urban species conservation has to take place in spite of urbanization, "in the midst of human enterprise" (Rosenzweig 2003). Every habitat within a city, no matter how semi-natural it is, is influenced by the urban environment. This restriction to biodiversity in urbanized areas cannot be eliminated but it can be buffered. Although urban species richness is restricted in the ways listed above, species conservation in cities does not have to be conservation of common or invasive species. The structural and geological heterogeneity of cities is one reason for their high species richness, and also functional and phylogenetic diversity can take advantage of this heterogeneity: On the one hand, semi-natural areas within cities, such as protected areas but also parks and gardens, can provide habitats for (rare) species with traits less suitable for the typical urbanized environment. On the other hand, typical urban-industrial habitats, such as urban brownfields, railway sites, harbor facilities, roadsides or walls provide habitats for species with traits tolerant to urban conditions and also for some rare species that lost their original habitat in non-urban landscapes (Lenzin *et al.* 2007). Both semi-natural protected areas and typical urban-industrial habitats together should allow a mix of semi-natural and typical urban floristic elements and therefore increase the functional and phylogenetic diversity of plant assemblages in urbanized areas. Moreover, the protection of rare species can especially contribute to the increase of phylogenetic diversity, because rare species are phylogenetically more diverse than common species (Ricotta *et al.* 2008b).

For such a "two-handed strategy" with semi-natural and typical urban habitats, urban-industrial habitats should be kept open for spontaneous vegetation, e.g. by unsealing of surfaces or not sealing unsealed surfaces, by not planting vegetation or by avoiding the use of pesticides and fertilizers (Wittig 1998; Lenzin *et al.* 2007). Intervention in protected areas should be restricted to actions necessary for maintaining a special state of succession, for protecting selected (rare) species or for safety of people. Protected areas in cities should also be connected with similar habitats in the city's environs. Further studies on functional and phylogenetic diversity of different types of habitats occurring in cities can provide valuable additional information on which habitats are especially able to support a functionally and phylogenetically diverse flora.

Conserving high species diversity in urbanized areas is not only valuable for species; it is also valuable for people: Nature in cities provides nearby possibilities for recreation and it contributes to environmental education. Because more than half of the world's total population lives in cities, nature in cities is for many people the only nature they experience everyday. We need to raise the awareness of people towards the importance of biodiversity and this is only possible if urban dwellers are involved and can experience biodiversity. "The battle for live on earth will be lost or won in cities" (Müller *et al.* 2008). A diverse urban flora helps to reach this target.

References

Ackerly D.D. & Donoghue M.J. (1995) Phylogeny and ecology reconsidered. *Journal of Ecology,* 83, 730-733.

Ackerly D.D. & Nyffeler R. (2004) Evolutionary diversification of continuous traits: phylogenetic tests and application to seed size in the California flora. *Evolutionary Ecology,* 18, 249-272.

Aitchison J. (1982) The Statistical Analysis of Compositional Data. *Journal of the Royal Statistical Society B,* 44, 139-177.

Antrop M. (2004) Landscape change and the urbanization process in Europe. *Landscape and Urban Planning,* 67, 9-26.

Araújo M.B. (2003) The coincidence of people and biodiversity in Europe. *Global Ecology and Biogeography,* 12, 5-12.

Badeck F.W., Bondeau A., Bottcher K., Doktor D., Lucht W., Schaber J. & Sitch S. (2004) Responses of spring phenology to climate change. *New Phytologist,* 162, 295-309.

Badeck F.W., Pompe S., Kühn I. & Glauer A. (2008) Zeitlich hochauflösende Klimainformationen auf dem Messtischblattraster und für Schutzgebiete in Deutschland. *Naturschutz und Landschaftsplanung,* 40, 343-345.

Balmford A., Moore J.L., Brooks T., Burgess N., Hansen L.A., Williams P. & Rahbek C. (2001) Conservation conflicts across Africa. *Science,* 291, 2616-2619.

Bañuelos M.J., Kollmann J., Hartvig P. & Quevedo M. (2004) Modelling the distribution of Ilex aquifolium at the north-eastern edge of its geographical range. *Nordic Journal of Botany,* 23, 129-142.

Barthlott W., Biedinger.N., Braun G., Feig F., Kier G. & Mutke J. (1999) Terminological and methodological aspects of the mapping and analysis of the global biodiversity. *Acta Botanica Fennica,* 162, 103-110.

Bekker R.M. & Kwak M.M. (2005) Life history traits as predictors of plant rarity, with particular reference to hemiparasitic Orobanchaceae. *Folia Geobotanica,* 40, 231-242.

Berger S., Sohlke G., Walther G.R. & Pott R. (2007) Bioclimatic limits and range shifts of cold-hardy evergreen broad-leaved species at their northern distributional limit in Europe. *Phytocoenologia,* 37, 523-539.

Berry B.J.L. (1990) Urbanization. In: *The earth as transformed by human action* (eds. Turner B.L.I., Clark W.C., Kates R.W., Richards J.F., Mathews J.T. & Meyer W.B.). Cambridge University Press, Cambridge, pp. 103-119.

Biesmeijer J.C., Roberts S.P.M., Reemer M., Ohlemüller R., Edwards M., Peeters T., Schaffers A.P., Potts S.G., Kleukers R., Thomas C.D., Settele J. & Kunin W.E. (2006) Parallel Declines in Pollinators and Insect-Pollinated Plants in Britain and the Netherlands. *Science,* 313, 351-354.

Billheimer D., Guttorp P. & Fagan F. (2001) Statistical Interpretation of Species Composition. *Journal of the American Statistical Association,* 96, 1205-1214.

Bivand R., Anselin L., Berke O., Bernat A., Carvalho M., Chun Y., Dormann C.F., Dray S., Halbersma R., Lewin-Koh N., Ma J., Millo G., Mueller W., Ono H., Peres-Neto P., Reder M., Tiefelsdorf M. & Yu D. (2007) *spdep: Spatial dependence: weighting schemes, statistics and models.* R package version 0.4-9.

Blair R.B. (1996) Land use and avian species diversity along an urban gradient. *Ecological Applications,* 6, 506-519.

Blume H.-P. (1998) Böden. In: *Stadtökologie. Ein Fachbuch für Studium und Praxis* (eds. Sukopp H. & Wittig R.). Gustav Fischer, Stuttgart, pp. 168-185.

Brandes D. (1993) Eisenbahnanlagen als Untersuchungsgegenstand der Geobotanik. *Tuexenia,* 13, 415-444.

Brandes D. & Oppermann F. (1995) Straßen, Kanäle und Bahnanlagen als lineare Strukturen in der Landschaft sowie deren Bedeutung für die Vegetation. *Berichte der Reihnhold-Tüxen-Gesellschaft,* 7, 89-110.

Briemle G., Nitsche S. & Nitsche L. (2002) Nutzungswertzahlen für Gefäßpflanzen des Grünlandes. In: *BiolFlor – Eine Datenbank mit biologisch-ökologischen Merkmalen zur Flora von Deutschland* (eds. Klotz S., Kühn I. & Durka W.). Bundesamt für Naturschutz, Bonn-Bad Godesberg, pp. 203-225.

Brown J.H. (1995) *Macroecology.* University of Chicago Press, Chicago.

Bundesanstalt für Geowissenschaften und Rohstoffe (1993) *Geologische Karte der Bundesrepublik Deutschland 1: 1,000,000.* Bundesanstalt für Geowissenschaften und Rohstoffe, Hannover.

Bundesanstalt für Geowissenschaften und Rohstoffe (1995) *Bodenübersichtskarte 1: 1,000,000.* Bundesanstalt für Geowissenschaften und Rohstoffe, Hannover.

Buschendorf J. & Klotz S. (1996) *Geschützte Natur in Halle (Saale). Flora und Fauna der Schutzgebiete. Teil 2: Flora der Schutzgebiete.* Umweltamt, Halle (Saale).

Buxbaum J.C. (1721) *Enumeratio plantarum accuratior in agro Hallensi locisque vicinis crescentium una cum aerum characteribus et viribus qua variae quam ante descriptae exhibentur.* Halae Magdeburgicae.

Cadotte M.W., Murray B.R. & Lovett-Doust J. (2006) Evolutionary and ecological influences of plant invader success in the flora of Ontario. *Ecoscience*, 13, 388-395.

Cardillo M., Purvis A., Sechrest W., Gittleman J.L., Bielby J. & Mace G.M. (2004) Human population density and extinction risk in the world's carnivores. *Plos Biology*, 2, 909-914.

Cavender-Bares J., Keen A. & Miles B. (2006) Phylogenetic structure of floridian plant communities depends on taxonomic and spatial scale. *Ecology*, 87, S109-S122.

Chazdon R.L., Careaga S., Webb C. & Vargas O. (2003) Community and phylogenetic structure of reproductive traits of woody species in wet tropical forests. *Ecological Monographs*, 73, 331-348.

Chesson P. (2000) Mechanisms of maintenance of species diversity. *Annual Review of Ecology and Systematics*, 31, 343-366.

Chocholoušková Z. & Pyšek P. (2003) Changes in composition and structure of urban flora over 120 years: a case study of the city of Plzeň. *Flora*, 198, 366-376.

Cincotta R.P., Wisnewski J. & Engelman R. (2000) Human population in the biodiversity hotspots. *Nature*, 404, 990-992.

Cornelissen J.H.C., Lavorel S., Garnier E., Diaz S., Buchmann N., Gurvich D.E., Reich P.B., ter Steege H., Morgan H.D., van der Heijden M.G.A., Pausas J.G. & Poorter H. (2003) A handbook of protocols for standardised and easy measurement of plant functional traits worldwide. *Australian Journal of Botany*, 51, 335-380.

Crawley M.J. (2007) *The R Book*. Wiley, Chichester.

Crawley M.J., Harvey P.H. & Purvis A. (1996) Comparative ecology of the native and alien floras of the British Isles. *Philosophical Transactions of the Royal Society B*, 351, 1251-1259.

Culley T.M., Weller S.G. & Sakai A.K. (2002) The evolution of wind pollination in angiosperms. *Trends in Ecology & Evolution*, 17, 361-369.

Díaz S. & Cabido M. (1997) Plant functional types and ecosystem function in relation to global change. *Journal of Vegetation Science*, 8, 463-474.

Díaz S., Cabido M., Zak M., Carretero E.M. & Aranibar J. (1999) Plant functional traits, ecosystem structure and land-use history along a climatic gradient in central-western Argentina. *Journal of Vegetation Science*, 10, 651-660.

Dobson A.P., Rodriguez J.P. & Roberts W.M. (2001) Synoptic tinkering: Integrating strategies for large-scale conservation. *Ecological Applications*, 11, 1019-1026.

Dobson F.S., Yu J.P. & Smith A.T. (1995) The importance of evaluating rarity. *Conservation Biology*, 9, 1648-1651.

Dony J.G. & Denholm I. (1985) Some Quantitative Methods of Assessing the Conservation Value of Ecologically Similar Sites. *Journal of Applied Ecology*, 22, 229-238.

Dormann C.F., McPherson J.M., Araujo M.B., Bivand R., Bolliger J., Carl G., Davies R.G., Hirzel A., Jetz W., Kissling W.D., Kühn I., Ohlemüller R., Peres-Neto P.R., Reineking B., Schröder B., Schurr F.M. & Wilson R. (2007) Methods to account for spatial autocorrelation in the analysis of species distributional data: a review. *Ecography*, 30, 609-628.

Düring C. (2004) Flora und Vegetation der Bahn- und Hafenanlagen im Großraum Regensburg. Typisierung der aktuellen Gefäßpflanzenflora, alpha-Diversität, Verbreitungsmuster, Pflanzengesellschaften. *Hoppea*, 65, 71-293.

Durka W. (2002) Phylogenie der Farn- und Blütenpflanzen Deutschlands. In: *BiolFlor – Eine Datenbank mit biologisch-ökologischen Merkmalen zur Flora von Deutschland* (eds. Klotz S., Kühn I. & Durka W.). Bundesamt für Naturschutz, Bonn-Bad Godesberg, pp. 75-91.

Ecke F. & Rydin H. (2000) Succession on a land uplift coast in relation to plant strategy theory. *Annales Botanici Fennici*, 37, 163-171.

Ehrlén J. & Eriksson O. (2000) Dispersal limitation and patch occupancy in forest herbs. *Ecology*, 81, 1667-1674.

Ellenberg H., Weber H.E., Düll R., Wirth V. & Werner W. (2001) Zeigerwerte von Pflanzen in Mitteleuropa. *Scripta Geobotanica*, 18, 1-262.

Elston D.A., Illius A.W. & Gordon I.J. (1996) Assessment of preference among a range of options using log ratio analysis. *Ecology*, 77, 2538-2548.

Environmental Agency Saxony-Anhalt (1997) *Soil map of Leipzig and surroundings, 1:50,000.* Landesamt für Geologie und Bergwesen, Halle (Saale).

Eversham B.C., Roy D.B. & Telfer M.G. (1996) Urban, industrial and other manmade sites as analogues of natural habitats for Carabidae. *Annales Zoologici Fennici*, 33, 149-156.

Faith D.P. (1992) Conservation evaluation and phylogenetic diversity. *Biological Conservation*, 61, 1-10.

Fitting H., Schulz A. & Wüst E. (1899) Nachtrag zu August Garckes Flora von Halle. *Verhandlungen des Botanischen Vereins der Provinz Brandenburg*, 41, 118-165.

Fitting H., Schulz A. & Wüst E. (1901) Nachtrag zu August Garckes Flora von Halle. *Verhandlungen des Botanischen Vereins der Provinz Brandenburg*, 43, 34-53.

Fitting H., Schulz A. & Wüst E. (1903) Beiträge zur Kenntnis der Flora der Umgebung von Halle a. S. *Zeitschrift für Naturwissenschaften*, 76, 110-113.

Fjeldså J. & Rahbek C. (1998) Continent-wide conservation priorities and diversification processes. In: *Conservation in a changing world* (eds. Mace G.M.,

Balmford A. & Ginsberg J.). Cambridge University Press, Cambridge, pp. 139-160.

Flynn S., Turner R.M. & Dickie J.B. (2004) Seed Information Database (release 6.0, Oct. 2004). http://www.kew.org/data/sid.

Franzaring J. & Fangmeier A. (2006) Approaches to the monitoring of atmospheric nitrogen using bioindicator plants. *Gefahrstoffe Reinhaltung der Luft*, 66, 253-259.

Fréville H., McConway K., Dodd M. & Silvertown J. (2007) Prediction of extinction in plants: Interaction of extrinsic threats and life history traits. *Ecology*, 88, 2662-2672.

Friedrich K. (2006) Die Stadt Halle und ihre Region. In: *Halle und sein Umland. Geographischer Exkursionsführer* (eds. Friedrich K. & Frühauf M.). Mitteldeutscher Verlag, Halle (Saale), pp. 9-15.

Fry J.M., Fry T.R.L. & McLaren K.R. (2000) Compositional data analysis and zeros in micro data. *Applied Economics*, 32, 953-959.

Garcke A. (1848) *Flora von Halle mit näherer Berücksichtigung der Umgegend von Weissenfels, Naumburg, Freiburg, Bibra, Nebra, Querfurt, Allstedt, Artern, Eisleben, Hettstedt, Sandersleben, Aschersleben, Stassfurt, Bernburg, Köthen, Dessau, Oranienbaum, Bitterfeld und Delitzsch. Erster Theil.* Halle (Saale).

Garcke A. (1856) *Flora von Halle mit näherer Berücksichtigung der Umgegend von Weissenfels, Naumburg, Freiburg, Bibra, Nebra, Querfurt, Allstedt, Artern, Eisleben, Hettstedt, Sandersleben, Aschersleben, Stassfurt, Bernburg, Köthen, Dessau, Oranienbaum, Bitterfeld und Delitzsch. Zweiter Theil. Kryptogamen nebst einem Nachtrage zu den Phanerogamen.* Halle (Saale).

Gaston K.J. (1994) *Rarity.* Chapman and Hall, London.

Gaston K.J. & Blackburn T.M. (2000) *Pattern and Process in Macroecology.* Blackwell Science, Oxford.

Gilbert O.L. (1989) *The Ecology of Urban Habitats.* Chapman and Hall, London.

Gloor S., Bontadina F., Hegglin D., Deplazes P. & Breitenmoser U. (2001) The rise of urban fox populations in Switzerland. *Mammalian Biology*, 66, 155-164.

Godefroid S. (2001) Temporal analysis of the Brussels flora as indicator for changing environmental quality. *Landscape and Urban Planning*, 52, 203-224.

Godefroid S. & Koedam N. (2007) Urban plant species patterns are highly driven by density and function of built-up areas. *Landscape Ecology*, 22, 1227-1239.

Gorelick R. (2001) Did insect pollination cause increased seed plant diversity? *Biological Journal of the Linnean Society,* 74, 407-427.

Grosse E. (1978) Neufunde und Bestätigungen aus dem Gebiet nördlich von Halle (Saale). *Mitteilungen zur floristischen Kartierung,* 4, 50-52.

Grosse E. (1979) Neufunde und Bestätigungen aus dem Gebiet nördlich von Halle (Saale) 2. Beitrag. *Mitteilungen zur floristischen Kartierung,* 5, 75-81.

Grosse E. (1981) Neufunde und Bestätigungen aus dem Gebiet nördlich von Halle (Saale) 3. Beitrag. *Mitteilungen zur floristischen Kartierung,* 7, 101-111.

Grosse E. (1983) *Anthropogene Florenveränderungen in der Agrarlandschaft nördlich von Halle (S.).* PhD thesis, Martin-Luther-University Halle-Wittenberg, Halle (Saale).

Grosse E. (1985) Beiträge zur Geschichte der Wälder des Stadtkreises Halle und des nördlichen Saalkreises. *Hercynia N. F.,* 22, 37-52.

Grosse E. (1987) Anthropogene Florenveränderungen in der Agrarlandschaft nördlich von Halle (Saale). 2. Folge: arten naturnaher Wälder. *Hercynia N. F.,* 24, 179-209.

Grosse E. & John H. (1987) Zur Flora von Halle und Umgebung. 1. Beitrag. *Mitteilungen zur floristischen Kartierung,* 13, 85-112.

Grosse E. & John H. (1989) Zur Flora von Halle und Umgebung. 2. Beitrag. *Mitteilungen zur floristischen Kartierung,* 15, 13-36.

Grosse E. & John H. (1991) Zur Flora von Halle und Umgebung. 3. Beitrag. *Mitteilungen zur floristischen Kartierung,* 17, 15-22.

Guisan A. & Thuiller W. (2005) Predicting species distribution: offering more than simple habitat models. *Ecology Letters,* 8, 993-1009.

Haeupler H. (1975) Statistische Auswertungen von Punktrasterkarten der Gefäßpflanzenflora Süd-Niedersachsens. *Scripta Geobotanica,* 8, 1-141.

Hampe A. (2004) Bioclimate envelope models: what they detect and what they hide. *Global Ecology and Biogeography,* 13, 469-471.

Harvey P.H. & Pagel M.D. (1991) *The Comparative Method in Evolutionary Biology.* Oxford University Press, Oxford.

Helmus M.R., Savage K., Diebel M.W., Maxted J.T. & Ives A.R. (2007) Separating the determinants of phylogenetic community structure. *Ecology Letters,* 10, 917-925.

Hodkinson D.J. & Thompson K. (1997) Plant dispersal: the role of man. *Journal of Applied Ecology,* 34, 1484-1496.

Hope D., Gries C., Zhu W.X., Fagan W.F., Redman C.L., Grimm N.B., Nelson A.L., Martin C. & Kinzig A. (2003) Socioeconomics drive urban plant diversity. *Proceedings of the National Academy of Sciences of the United States of America,* 100, 8788-8792.

Hulme P.E. (2008) Contrasting alien and native plant species-area relationships: the importance of spatial grain and extent. *Global Ecology and Biogeography,* 17, 641-647.

Institut für Länderkunde Leipzig (2003) *Nationalatlas Bundesrepublik Deutschland. Relief, Boden und Wasser.* Spektrum Akademischer Verlag, Heidelberg, Berlin.

Intergovernmental Panel on Climate Change (IPCC) (2007) *Climate Change 2007: Synthesis Report.* Intergovernmental Panel on Climate Change, Geneva, Switzerland.

Kasanko M., Barredo J.I., Lavalle C., McCormick N., Demicheli L., Sagris V. & Brezger A. (2006) Are European cities becoming dispersed? A comparative analysis of 15 European areas. *Landscape and Urban Planning,* 77, 111-130.

Kazakou E., Vile D., Shipley B., Gallet C. & Garnier E. (2006) Co-variations in litter decomposition, leaf traits and plant growth in species from a Mediterranean old-field succession. *Functional Ecology,* 20, 21-30.

Kent M., Stevens R.A. & Zhang L. (1999) Urban plant ecology patterns and processes: a case study of the flora of the City of Plymouth, Devon, UK. *Journal of Biogeography,* 26, 1281-1298.

Kissling W.D. & Carl G. (2008) Spatial autocorrelation and the selection of simultaneous autoregressive models. *Global Ecology and Biogeography,* 17, 59-71.

Klein A.M., Vaissiere B.E., Cane J.H., Steffan-Dewenter I., Cunningham S.A., Kremen C. & Tscharntke T. (2007) Importance of pollinators in changing landscapes for world crops. *Proceedings of the Royal Society B-Biological Sciences,* 274, 303-313.

Kleyer M. (1999) Distribution of plant functional types along gradients of disturbance intensity and resource supply in an agricultural landscape. *Journal of Vegetation Science,* 10, 697-708.

Kleyer M. (2002) Validation of plant functional types across two contrasting landscapes. *Journal of Vegetation Science,* 13, 167-178.

Kleyer M., Bekker R.M., Knevel I.C., Bakker J.P., Thompson K., Sonnenschein M., Poschlod P., van Groenendael J.M., Klimeš L., Klimešová J., Klotz S., Rusch G.M., Hermy M., Adriaens D., Boedeltje G., Bossuyt B., Dannemann A., Endels P., Götzenberger L., Hodgson J.G., Jackel A.-K., Kühn I., Kunzmann D., Ozinga W.A., Römermann C., Stadler M., Schlegelmilch J., Steendam H.J., Tackenberg O., Wilmann B., Cornelissen J.H.C., Eriksson O., Garnier E. & Peco B. (2008) The LEDA Traitbase: A database of life-history traits of Northwest European flora. *Journal of Ecology,* 96, 1266-1274.

Klotz S. (1984) *Phytoökologische Beiträge zur Charakterisierung und Gliederung urbaner Ökosysteme, dargestellt am Beispiel der Städte Halle und Halle-Neustadt.* PhD thesis, Martin-Luther-University Halle-Wittenberg.

Klotz S. (1989) Merkmale der Stadtflora. *Braun-Blanquetia,* 3, 57-60.

Klotz S. & Stolle J. (1998) Farn- und Blütenpflanzen. In: *Arten- und Biotopschutzprogramm Sachsen-Anhalt* Berichte des Landesamtes für Umweltschutz, special edition 4/1998, Halle (Saale).

Klotz S., Kühn I. & Durka W. (2002) BiolFlor – Eine Datenbank mit biologischökologischen Merkmalen zur Flora von Deutschland. *Schriftenreihe für Vegetationskunde,* 38, 1-333.

Klötzli F. & Walther G.R., eds. (1999) *Conference on recent shifts in vegetation boundaries of deciduous forests, especially due to general global warming.* Birkhäuser, Basel.

Knapp R. (1944a) *Vegetationsaufnahmen von Trockenrasen und Felsfluren Mitteldeutschlands, Teil 2: Atlantisch-Submediterrane und Dealpine Trockenrasen (Bromion erecti).* Halle (Saale).

Knapp R. (1944b) *Über Zwergstrauch-Heiden im Mitteldeutschen Trocken-Gebiet.* Halle (Saale).

Knapp R. (1945) *Die Ruderalgesellschaften in Halle an der Saale und seiner Umgebung.* Halle (Saale).

Knapp S., Kühn I., Mosbrugger V. & Klotz S. (2008a) Do protected areas in urban and rural landscapes differ in species diversity? *Biodiversity and Conservation,* 17, 1595-1612.

Knapp S., Kühn I., Wittig R., Ozinga W.A., Poschlod P. & Klotz S. (2008b) Urbanization causes shifts of species' trait state frequencies. *Preslia,* 80, 375-388.

Knauth C. (1687) *Enumeratio plantarum circa Halam Saxonum et in ejus vicinia, ad trium fere miliarium serptium, sponte provenientium.* Lipsiae.

Knevel I.C., Bekker R.M., Kunzmann D., Stadler M. & Thompson K. (2005) *The LEDA Traitbase Collecting and Measuring Standards of Life-history Traits of the Northwest European Flora.* LEDA Traitbase project, http://www.leda-traitbase.org.

Kollmann J. (1994) Ausbreitungsökologie endozoochorer Gehölzarten. Naturschutzorientierte Untersuchungen über die Rolle von Gehölzen bei der Erhaltung, Entwicklung und Vernetzung von Ökosystemen. *Veröffentlichungen Projekt angewandte Ökologie,* 9, 1-212.

Korneck D., Schnittler M., Klingenstein F., Ludwig G., Takla M., Bohn U. & May R. (1998) Warum verarmt unsere Flora? Auswertung der Roten Liste der Farn- und Blütenpflanzen Deutschlands. *Schriftenreihe für Vegetationskunde,* 29, 299-444.

Kowarik I. (2008) On the role of alien species in urban flora and vegetation. In: *Urban Ecology. An International Perspective on the Interaction Between Humans and Nature* (eds. Marzluff J.M., Shulenberger E., Endlicher W., Alberti M., Bradley G., Ryan C., Simon U. & ZumBrunnen C.). Springer, New York, pp. 321-338.

Kowarik I. & Säumel I. (2007) Biological flora of Central Europe: Ailanthus altissima (Mill.) Swingle. *Perspectives in Plant Ecology Evolution and Systematics,* 8, 207-237.

Krause A. (1998) Floras Alltagskleid oder Deutschlands 100 häufigste Pflanzenarten. *Natur und Landschaft,* 73, 486-491.

Kühn I. (2007) Incorporating spatial autocorrelation may invert observed patterns. *Diversity and Distributions,* 13, 66-69.

Kühn I. & Klotz S. (2002) Floristic status and alien species. In: *BiolFlor – Eine Datenbank mit biologisch-ökologischen Merkmalen zur Flora von Deutschland* (eds. Klotz S., Kühn I. & Durka W.). Bundesamt für Naturschutz, Bonn-Bad Godesberg, pp. 47-56.

Kühn I. & Klotz S. (2006) Urbanization and homogenization - Comparing the floras of urban and rural areas in Germany. *Biological Conservation,* 127, 292-300.

Kühn I. & Klotz S. (2007) From Ecosystem invasibility to local, regional and global patterns of invasive species. In: *Biological invasions. Ecological Studies 193* (ed. Nentwig W.). Springer, Berlin, Heidelberg, New York, pp. 181-196.

Kühn I., Brandl R., May R. & Klotz S. (2003) Plant distribution patterns in Germany – Will aliens match natives? *Feddes Repertorium,* 114, 559-573.

Kühn I., Brandl R. & Klotz S. (2004a) The flora of German cities is naturally species rich. *Evolutionary Ecology Research,* 6, 749-764.

Kühn I., Durka W. & Klotz S. (2004b) BiolFlor - a new plant-trait database as a tool for plant invasion ecology. *Diversity and Distributions,* 10, 363-365.

Kühn I., Bierman S.M., Durka W. & Klotz S. (2006) Relating geographical variation in pollination types to environmental and spatial factors using novel statistical methods. *New Phytologist,* 172, 127-139.

Kühn I., Böhning-Gaese K., Cramer W. & Klotz S. (2008) Macroecology meets global change research. *Global Ecology and Biogeography,* 17, 3-4.

Kunin W.E. (1998) Extrapolating species abundance across spatial scales. *Science,* 281, 1513-1515.

Kuttler W. (1993) Stadtklima. In: *Stadtökologie* (eds. Sukopp H. & Wittig R.). Gustav Fischer, Stuttgart, Jena, New York, pp. 113-153.

Kuttler W. (2008) The urban climate - basic and applied aspects. In: *Urban ecology. An international perspective on the interaction between humans and na-*

ture (eds. Marzluff J.M., Shulenberger E., Endlicher W., Alberti M., Bradley G., Ryan C., Simon U. & ZumBrunnen C.). Springer, New York, pp. 233-248.

Landesamt für Umweltschutz Sachsen-Anhalt (2005) *Datenbank der Farn- und Blütenpflanzen Sachsen-Anhalts.* Landesamt für Umweltschutz Sachsen-Anhalt, Halle (Saale).

Landsberg H. (1981) *The Urban Climate.* Academic Press, New York.

Lang G. (1994) *Quartäre Vegetationsgeschichte Europas: Methoden und Ergebnisse.* Gustav Fischer Verlag, Jena, Stuttgart, New York.

Lavergne S., Molina J. & Debussche M. (2006) Fingerprints of environmental change on the rare Mediterranean flora: a 115-year study. *Global Change Biology,* 12, 1466-1478.

Lavorel S. & Garnier E. (2002) Predicting changes in community composition and ecosystem functioning from plant traits: revisiting the Holy Grail. *Functional Ecology,* 16, 545-556.

Lenzin H., Meier-Küpfer H., Schwegler S. & Baur B. (2007) Hafen- und Gewerbegebiete als Schwerpunkte pflanzlicher Diversität innerhalb urban-industrieller Ökosysteme. *Naturschutz und Landschaftsplanung,* 39, 86-93.

Leysser F.W. (1761) *Flora Halensis exhibens plantas circa Halam Salicam tes secundum systema sexuale Linnaeanum distributas. Ed. I.* Halle (Saale).

Leysser F.W. (1783) *Flora Halensis exhibens plantas circa Halam Salicam tes secundum systema sexuale Linnaeanum distributas. Ed. II.* Halle (Saale).

Lososová Z., Chytrý M., Kühn I., Hájek O., Horáková V., Pyšek P. & Tichý L. (2006) Patterns of plant traits in annual vegetation of man-made habitats in central Europe. *Perspectives in Plant Ecology, Evolution and Systematics,* 8, 69-81.

Lu P.L., Yu Q., Liu J.D. & Lee X.H. (2006) Advance of tree-flowering dates in response to urban climate change. *Agricultural and Forest Meteorology,* 138, 120-131.

Luo Z.K., Sun O.J., Ge Q.S., Xu W.T. & Zheng J.Y. (2007) Phenological responses of plants to climate change in an urban environment. *Ecological Research,* 22, 507-514.

MacDonald D., Crabtree J.R., Wiesinger G., Dax T., Stamou N., Fleury P., Lazpita J.G. & Gibon A. (2000) Agricultural abandonment in mountain areas of Europe: Environmental consequences and policy response. *Journal of Environmental Management,* 59, 47-69.

Mac Nally R. (2000) Regression and model-building in conservation biology, biogeography and ecology: The distinction between and reconciliation of 'predictive' and 'explanatory' models. *Biodiversity and Conservation,* 9, 655-671.

Magurran A.E. (2004) *Measuring Biological Diversity*. Blackwell Publishing, Oxford.

Maherali H. & Klironomos J.N. (2007) Influence of phylogeny on fungal community assembly and ecosystem functioning. *Science*, 316, 1746-1748.

Martin-Fernandez J.A., Barcelo-Vidal C. & Pawlowsky-Glahn V. (2000) Zero replacement in compositional datasets. In: *Studies in classification, data analysis, and knowledge organization* (eds. Kiers H., Rasson J., Groenen P. & Shader M.). Springer, Berlin, pp. 155-160.

Matlack G.R. (2005) Slow plants in a fast forest: local dispersal as a predictor of species frequencies in a dynamic landscape. *Journal of Ecology*, 93, 50-59.

McDonnell M.J. (1997) A Paradigm Shift. *Urban Ecosystems*, 1, 85-86.

McDonnell M.J., Pickett S.T.A., Groffman P., Bohlen P., Pouyat R., Zipperer W.C., Parmelee R.W., Carreiro M.M. & Medley K. (1997) Ecosystem processes along an urban-to-rural gradient. *Urban Ecosystems*, 1, 21-36.

McKinney M.L. (2002) Do human activities raise species richness? Contrasting patterns in United States plants and fishes. *Global Ecology and Biogeography*, 11, 343-348.

McKinney M.L. (2006) Urbanization as a major cause of biotic homogenization. *Biological Conservation*, 127, 247-260.

Ministerium für Raumordnung, Landwirtschaft und Umwelt des Landes Sachsen-Anhalt (1996) *Agraratlas des Landes Sachsen-Anhalt. Landwirtschaftliches Gutachten in Karten, Texten, Übersichten*. Ministerium für Raumordnung,Landwirtschaft und Umwelt des Landes Sachsen-Anhalt, Magdeburg.

Müller N., Knight D. & Werner P. (2008) Cities and the Convention on Biological Biodiversity - from Rio via Curitiba to Erfurt - facing the main challenges of this century for life on earth. In: *Urban Biodiversity & Design - Third Conference of the Competence Network Urban Ecology Teil 1: Book of Abstracts* (eds. Müller N., Knight D. & Werner P.). BfN-Skripten 229-1, Bundesamt für Naturschutz, Bonn, pp. 9-9, http://www.bfn.de/0502_skripten.html.

Müller-Westermeier G., Kreis A. & Dittmann E. (1999) *Klimaatlas Bundesrepublik Deutschland. Teil 1. Lufttemperatur, Niederschlagshöhe, Sonnenscheindauer*. Deutscher Wetterdienst, Offenbach am Main.

Müller-Westermeier G., Kreis A. & Dittmann E. (2001) *Klimaatlas Bundesrepublik Deutschland. Teil 2. Verdunstung, Maximumtemperatur, Minimumtemperatur, Kontinentalität*. Deutscher Wetterdienst, Offenbach am Main.

Niemelä J. (1999) Is there a need for a theory of urban ecology? *Urban Ecosystems*, 3, 57-65.

Niinemets U. & Peñuelas J. (2008) Gardening and urban landscaping: Significant players in global change. *Trends in Plant Science*, 13, 60-65.

Oke T.R. (1982) The energetic basis of the urban heat-island. *Quarterly Journal of the Royal Meteorological Society*, 108, 1-24.

Ozinga W.A., Schaminee J.H.J., Bekker R.M., Bonn S., Poschlod P., Tackenberg O., Bakker J. & van Groenendael J.M. (2005) Predictability of plant species composition from environmental conditions is constrained by dispersal limitation. *Oikos*, 108, 555-561.

Paradis E., Strimmer K., Claude J., Jobb G., Opgen-Rhein R., Dutheil J., Noel Y., Bolker B. & Lemon J. (2006) *ape: Analyses of Phylogenetics and Evolution*. R package version 1.8-3.

Paul M.J. & Meyer J.L. (2001) Streams in the urban landscape. *Annual Review of Ecology and Systematics*, 32, 333-365.

Pautasso M. (2007) Scale dependence of the correlation between human population presence and vertebrate and plant species richness. *Ecology Letters*, 10, 16-24.

Pauw A. (2007) Collapse of a pollination web in small conservation areas. *Ecology*, 88, 1759-1769.

Pinheiro J., Bates D., DebRoy S. & Sarkar D. (2006) *nlme: Linear and nonlinear mixed effects models*. R package version 3.1-73.

Poschlod P. & Bonn S. (1998) Changing dispersal processes in the central European landscape since the last ice age: an explanation for the actual decrease of plant species richness in different habitats? *Acta Botanica Neerlandica*, 47, 27-44.

Poschlod P., Kleyer M. & Tackenberg O. (2000) Databases on life history traits as a tool for risk assessment in plant species. *Zeitschrift für Ökologie und Naturschutz*, 9, 3-18.

Pressey R.L. (1994) Ad Hoc Reservations - Forward Or Backward Steps in Developing Representative Reserve Systems. *Conservation Biology*, 8, 662-668.

Preston C.D. (2000) Engulfed by suburbia or destroyed by the plough: the ecology of extinction in Middlesex and Cambridgeshire. *Watsonia*, 23, 59-81.

Prinzing A., Durka W., Klotz S. & Brandl R. (2001) The niche of higher plants: evidence for phylogenetic conservatism. *Proceedings of the Royal Society B-Biological Sciences*, 268, 2383-2389.

Prinzing A., Reiffers R., Braakhekke W.G., Hennekens S.M., Tackenberg O., Ozinga W.A., Schaminée J.H.J. & van Groenendael J.M. (2008) Less lineages - more trait variation: phylogenetically clustered plant communities are functionally more diverse. *Ecology Letters*, 11, 809-819.

Purvis A. & Hector A. (2000) Getting the measure of biodiversity. *Nature*, 405, 212-219.

Pyšek P. (1993) Factors affecting the diversity of flora and vegetation in central European settlements. *Vegetatio*, 106, 89-100.

Pyšek P. (1995) Approaches to studying spontaneous settlement flora and vegetation in Central Europe: A review. In: *Urban Ecology as the Basis of Urban Planning* (eds. Sukopp H., Numata M. & Huber A.). The Hague: SPB Academic Publishing, pp. 23-29.

Pyšek P. (1998) Is there a taxonomic pattern to plant invasion? *Oikos,* 82, 282-294.

Pyšek P., Zdena C., Pyšek A., Jarosik V., Chytry M. & Tichy L. (2004) Trends in species diversity and composition of urban vegetation over three decades. *Journal of Vegetation Science,* 15, 781-788.

R Development Core Team (2006) *R: A Language and Environment for Statistical Computing.* R Foundation for Statistical Computing, Vienna.

R Development Core Team (2007) *R: A Language and Environment for Statistical Computing.* R Foundation for Statistical Computing, Vienna, Austria.

Rabinowitz D. (1981) Seven forms of rarity. In: *The biological aspects of rare plant conservation* (ed. Synge H.). Wiley and Sons, Chichester, pp. 205-217.

Rao C.R. (1982) Diversity and dissimilarity coefficients - a unified approach. *Theoretical Population Biology,* 21, 24-43.

Raschke W. & Schultz A. (2006) Stadtbevölkerung im Wandel - Die Bevölkerungsentwicklung und - struktur von Halle. In: *Halle und sein Umland. Geographischer Exkursionsführer* (eds. Friedrich K. & Frühauf M.). Mitteldeutscher Verlag, Halle (Saale), pp. 50-56.

Rauschert S. (1966a) Aufruf zur Neubestätigung verschollener und zweifelhafter Pflanzenfundorte im Bezirk Halle. *Wissenschaftliche Zeitschrift der Universität Halle,* 15, 774-778.

Rauschert S. (1966b) Zur Flora des Bezirkes Halle. *Wissenschaftliche Zeitschrift der Universität Halle,* 15, 737-750.

Rauschert S. (1967) Zur Flora des Bezirkes Halle (2. Beitrag). *Wissenschaftliche Zeitschrift der Universität Halle,* 16, 867-868.

Rauschert S. (1972) Zur Flora des Bezirkes Halle (4. Beitrag). *Wissenschaftliche Zeitschrift der Universität Halle,* 21, 63-65.

Rauschert S. (1973) Zur Flora des Bezirkes Halle (5. Beitrag). *Wissenschaftliche Zeitschrift der Universität Halle,* 22, 32-33.

Rauschert S. (1975) Zur Flora des Bezirkes Halle (6. Beitrag). *Wissenschaftliche Zeitschrift der Universität Halle,* 24, 84-91.

Rauschert S. (1977a) 400 Jahre "Sylva Hercynia" von Johannes Thal. *Hercynia N. F.,* 14, 361-374.

Rauschert S. (1977b) Zur Flora des Bezirkes Halle (7. Beitrag). *Mitteilungen zur floristischen Kartierung,* 3, 50-56.

Rauschert S. (1979) Zur Flora des Bezirkes Halle (8. Beitrag). *Mitteilungen zur floristischen Kartierung,* 5, 57-73.

Rauschert S. (1980) Zur Flora des Bezirkes Halle (9. Beitrag). *Mitteilungen zur floristischen Kartierung,* 6, 30-36.

Rauschert S. (1982) Zur Flora des Bezirkes Halle (10. Beitrag). *Mitteilungen zur floristischen Kartierung,* 8, 55-59.

Ricotta C., DiNepi M., Guglietta D. & Celesti-Grapow L. (2008a) Exploring taxonomic filtering in urban environments. *Journal of Vegetation Science,* 19, 229-238.

Ricotta C., Godefroid S. & Celesti-Grapow L. (2008b) Common species have lower taxonomic diversity - Evidence from the urban floras of Brussels and Rome. *Diversity and Distributions,* 14, 530-537.

Roche P., Díaz-Burlinson N. & Gachet S. (2004) Congruency analysis of species ranking based on leaf traits: which traits are the more reliable? *Plant Ecology,* 174, 37-48.

Roetzer T., Wittenzeller M., Haeckel H. & Nekovar J. (2000) Phenology in central Europe - differences and trends of spring phenophases in urban and rural areas. *International Journal of Biometeorology,* 44, 60-66.

Römermann C. (2006) *Patterns and processes of plant species frequency and life-history traits.* Gebrüder Bornträger, Berlin, Stuttgart, PhD thesis, University of Regensburg.

Römermann C., Tackenberg O., Scheuerer M., May R. & Poschlod P. (2007) Predicting habitat distribution and frequency from plant species co-occurrence data. *Journal of Biogeography,* 34, 1041-1052.

Römermann C., Tackenberg O., Jackel A.K. & Poschlod P. (2008) Eutrophication and fragmentation are related to species ' rate of decline but not to species rarity: results from a functional approach. *Biodiversity and Conservation,* 17, 591-604.

Rosenzweig M. (1995) *Species diversity in space and time.* Cambridge University Press, Cambridge.

Rosenzweig M. (2003) *Win-Win Ecology. How the Earth's Species Can Survive in the Midst of Human Enterprise.* Oxford University Press, New York.

Roth A. (1783) Additamenta ad Floram Halensam (edita a J.J. Reichhard). *Nova Acta Leopoldina,* 7, 201.

Sala O.E., Chapin F.S., Armesto J.J., Berlow E., Bloomfield J., Dirzo R., Huber-Sanwald E., Huenneke L.F., Jackson R.B., Kinzig A., Leemans R., Lodge D.M., Mooney H.A., Oesterheld M., Poff N.L., Sykes M.T., Walker B.H., Walker M. & Wall D.H. (2000) Biodiversity - Global biodiversity scenarios for the year 2100. *Science,* 287, 1770-1774.

Sax D.F. & Gaines S.D. (2003) Species diversity: from global decreases to local increases. *Trends in Ecology & Evolution,* 18, 561-566.

Schluter D. (2000) Ecological character displacement in adaptive radiation. *American Naturalist,* 156, S4-S16.

Schulz A. & Wüst E. (1906) Beiträge zur Kenntnis der Flora der Umgegend von Halle a. S. II. *Zeitschrift für Naturwissenschaften,* 78, 166-171.

Schulz A. & Wüst E. (1907) Beiträge zur Kenntnis der Flora der Umgegend von Halle a. S. III. *Zeitschrift für Naturwissenschaften,* 79, 267-271.

Schwartz M.W., Thorne J.H. & Viers J.H. (2006) Biotic homogenization of the California flora in urban and urbanizing regions. *Biological Conservation,* 127, 282-291.

Schweiger O., Maelfait J.P., Van Wingerden W., Hendrickx F., Billeter R., Speelmans M., Augenstein I., Aukema B., Aviron S., Bailey D., Bukacek R., Burel F., Diekotter T., Dirksen J., Frenzel M., Herzog F., Liira J., Roubalova M. & Bugter R. (2005) Quantifying the impact of environmental factors on arthropod communities in agricultural landscapes across organizational levels and spatial scales. *Journal of Applied Ecology,* 42, 1129-1139.

Schweiger O., Klotz S., Durka W. & Kühn I. (2008) A test of phylogenetic diversity indices. *Oecologia,* 157, 485-495.

Sechrest W., Brooks T.M., da Fonseca G.A.B., Konstant W.R., Mittermeier R.A., Purvis A., Rylands A.B. & Gittleman J.L. (2002) Hotspots and the conservation of evolutionary history. *Proceedings of the National Academy of Sciences of the United States of America,* 99, 2067-2071.

Smart S.M., Bunce R.G.H., Marrs R., Leduc M., Firbank L.G., Maskell L.C., Scott W.A., Thompson K. & Walker K.J. (2005) Large-scale changes in the abundance of common higher plant species across Britain between 1978, 1990 and 1998 as a consequence of human activity: Tests of hypothesised changes in trait representation. *Biological Conservation,* 124, 355-371.

Spilger L. (1937) Aus Senckenbergs botanischen Aufzeichnungen (1730/31) über Halle. *Hercynia,* 1, 166-173.

Sprengel C. (1806) *Florae, Halensis tentamen novum.* Halle (Saale).

Stadt Halle (2003a) *Digitale Karte der Schutzgebiete 1: 20 000.* Fachbereich Umwelt, Halle (Saale).

Stadt Halle (2003b) *Digitale Karte der flächendeckenden Biotoptypenkartierung 1:5 000.* Fachbereich Stadtentwicklung und Stadtplanung, Halle (Saale).

Statistisches Bundesamt (1997) *Daten zur Bodenbedeckung für die Bundesrepublik Deutschland 1: 100,000.* Statistisches Bundesamt, Wiesbaden.

Stolle J. & Klotz S. (2004) *Flora der Stadt Halle (Saale).* Calendula, hallesche Umweltblätter, Halle (Saale).

Sudnik-Wójcikowska B. & Galera H. (2005) Floristic differences in some anthropogenic habitats in Warsaw. *Annales Botanici Fennici,* 42, 185-193.

Sukopp H. (1998) Urban Ecology - Scientific and Practical Aspects. In: *Urban Ecology* (eds. Breuste J., Feldmann H. & Uhlmann O.). Springer Verlag, Berlin Heidelberg, pp. 3-16.

Sukopp H. & Starfinger U. (1999) Disturbance in urban ecosystems. In: *Ecosystems of disturbed ground* (ed. Walker L.R.). Elsevier, Amsterdam, pp. 397-412.

Sukopp H., Blume H.-P. & Kunick W. (1979) The soil, flora, and vegetation of Berlin's waste lands. In: *Nature in cities: The natural environment in the design and development of urban green space* (ed. Laurie I.C.). Wiley, Chichester, pp. 115-132.

Swenson N.G., Enquist B.J., Pither J., Thompson J. & Zimmerman J.K. (2006) The problem and promise of scale dependency in community phylogenetics. *Ecology,* 87, 2418-2424.

Swenson N.G., Enquist B.J., Thompson J. & Zimmerman J.K. (2007) The influence of spatial and size scale on phylogenetic relatedness in tropical forest communities. *Ecology,* 88, 1770-1780.

Tait C.J., Daniels C.B. & Hill R.S. (2005) Changes in species assemblages within the Adelaide Metropolitan Area, Australia, 1836-2002. *Ecological Applications,* 15, 346-359.

Tamis W.L.M., Van't Zelfde M., Van der Meijden R. & De Haes H.A.U. (2005) Changes in vascular plant biodiversity in the Netherlands in the 20th century explained by their climatic and other environmental characteristics. *Climatic Change,* 72, 37-56.

Ter Braak C.J.F. & Barendregt L.G. (1986) Weighted Averaging of Species Indicator Values - Its Efficiency in Environmental Calibration. *Mathematical Biosciences,* 78, 57-72.

Thal J. (1977) *Sylva Hercynia: sive catalogus plantarum sponte nascentium in montibus & locis plerisque Hercyniae Sylvae quae respicit Saxoniam. Edited, translated into German, interpreted and explained by Rauschert, S.* Zentralantiquariat der Deutschen Demokratischen Republik, Leipzig.

Thompson K. & McCarthy M.A. (2008) Traits of British alien and native urban plants. *Journal of Ecology,* 96, 853-859.

Thuiller W. (2003) BIOMOD - optimizing predictions of species distributions and projecting potential future shifts under global change. *Global Change Biology,* 9, 1353-1362.

Thuiller W., Araujo M.B. & Lavorel S. (2003) Generalized models vs. classification tree analysis: Predicting spatial distributions of plant species at different scales. *Journal of Vegetation Science,* 14, 669-680.

Tremlová K. & Münzbergová Z. (2007) Importance of species traits for species distribution in fragmented landscapes. *Ecology,* 88, 965-977.

Umweltbundesamt, Statistisches Bundesamt & Bundesanstalt für Geowissenschaften und Rohstoffe (2007) *Umweltdaten Deutschland. Nachhaltig wirtschaften - Natürliche Ressourcen und Umwelt schonen.* Umweltbundesamt, Dessau.

United Nations (2003) *Population, Education and Development. The Concise Report.* United Nations, New York, www.un.org/esa/population/publications/concise2003/Concisereport2003.pdf (accessed on 03rd of December 2007).

United Nations (2006) *World Urbanization Prospects. The 2005 Revision. Executive Summary, Fact Sheets, Data Tables.* United Nations, New York, www.un.org/esa/population/publications/WUP2005/2005WUPHighlights_Final_Report.pdf (accessed on 24th of July 2007).

United Nations (2008) *World Urbanization Prospects. The 2007 Revision. Highlights.* United Nations, New York, http://www. un.org/esa/population/ publications/wup2007/2007WUP_Highlights_web.pdf (accessed on 26th of March 2008).

Van der Veken S., Verheyen K. & Hermy M. (2004) Plant species loss in an urban area (Turnhout, Belgium) from 1880 to 1999 and its environmental determinants. *Flora,* 199, 516-523.

Vane-Wright R.I., Humphries C.J. & Williams P.H. (1991) What to protect? - Systematics and the agony of choice. *Biological Conservation,* 55, 235-254.

Vitousek P.M., Mooney H.A., Lubchenco J. & Melillo J.M. (1997) Human domination of Earth's ecosystems. *Science,* 277, 494-499.

von der Lippe M. & Kowarik I. (2007) Long-distance dispersal of plants by vehicles as a driver of plant invasions. *Conservation Biology,* 21, 986-996.

von der Lippe M. & Kowarik I. (2008) Do cities export biodiversity? Traffic as dispersal vector across urban-rural gradients. *Diversity and Distributions,* 14, 18-25.

Wagenbreth O. & Steiner W. (1982) *Geologische Streifzüge - Landschaft und Erdgeschichte zwischen Kap Arkona und Fichtelberg.* VEB Deutscher Verlag für Grundstoffindustrie, Leipzig.

Wallroth C.F.W. (1815) *Annus botanicus sive supplementum tertium ad Curtii Sprenglii Floram Halensem.* Halle (Saale).

Wallroth C.F.W. (1822) *Schedulae criticae de plantis florae Halensis selectis. Corollarium novum ad C.Sprengelii floram Halensem. Accedunt generum quorumdam specierumque omnium definitiones novae, excursus in stirpes difficiliores et icones V. Tom. I. Phanerogamia.* Halle (Saale).

Walossek W. (2006) 1200 Jahre Halle - Zur Grundrissentwicklung der Stadt. In: *Halle und sein Umland. Geographischer Exkursionsführer* (eds. Friedrich K. & Frühauf M.). Mitteldeutscher Verlag, Halle (Saale), pp. 33-41.

Walossek W., Winde F. & Zinke G. (2006) "An der Saale hellem Strande?" - ein Flussschicksal in Vergangenheit und Gegenwart. In: *Halle und sein Umland. Geographischer Exkursionsführer* (eds. Friedrich K. & Frühauf M.). Mitteldeutscher Verlag, Halle (Saale), pp. 66-75.

Walters S.M. (1970) The next twenty years. In: *The flora of a Changing Britain* (ed. Perring F.). Classey, Hampton, pp. 136-141.

Walther G.R., Gritti E.S., Berger S., Hickler T., Tang Z.Y. & Sykes M.T. (2007) Palms tracking climate change. *Global Ecology and Biogeography,* 16, 801-809.

Wangerin W. (1909) Die Vegetationsverhältnisse. In: *Heimatkunde des Saalkreises einschließlich des Stadtkreises Halle und des Mansfelder Seekreises* (ed. Ule W.). Verlag der Buchhandlung des Waisenhauses, Halle (Saale).

Wania A., Kühn I. & Klotz S. (2006) Plant richness patterns in agricultural and urban landscapes in Central Germany—spatial gradients of species richness. *Landscape and Urban Planning,* 75, 97-110.

Wardle D.A., Bardgett R.D., Klironomos J.N., Setala H., van der Putten W.H. & Wall D.H. (2004) Ecological linkages between aboveground and belowground biota. *Science,* 304, 1629-1633.

Warwick R.M. & Clarke K.R. (1998) Taxonomic distinctness and environmental assessment. *Journal of Applied Ecology,* 35, 532-543.

Webb C.O., Ackerly D.D., McPeek M.A. & Donoghue M.J. (2002) Phylogenies and community ecology. *Annual Review of Ecology and Systematics,* 33, 475-505.

Wessolek G. & Renger M. (1998) Bodenwasser. und Grundwasserhaushalt. In: *Stadtökologie. Ein Fachbuch für Studium und Praxis* (eds. Sukopp H. & Wittig R.). Gustav Fischer, Stuttgart, pp. 186-200.

Williams N.S.G., Morgan J.W., McDonnell M.J. & McCarthy M.A. (2005) Plant traits and local extinctions in natural grasslands along an urban-rural gradient. *Journal of Ecology,* 93, 1203-1213.

Williamson M.H. & Fitter A. (1996) The characters of successful invaders. *Biological Conservation,* 78, 163-170.

Winter M., Kühn I., Nentwig W. & Klotz S. (2008) Spatial aspects of trait homogenization within the German Flora. *Journal of Biogeography,* 35, 2289-2297.

Wittig R. (1998) Flora und Vegetation. In: *Stadtökologie. Ein Fachbuch für Studium und Praxis* (eds. Sukopp H. & Wittig R.). Gustav Fischer, Stuttgart, pp. 219-265.

Wittig R. (2002) *Siedlungsvegetation.* Ulmer, Stuttgart.

Wittig R. & Durwen K.-J. (1982) Ecological indicator-value spectra of spontaneous urban floras. In: *Urban ecology* (eds. Bornkamm R., Lee J.A. & Seaward M.R.D.). Blackwell Scientific Publications, Oxford, pp. 23-31.

Wittig R. & Ou X. (1993) Analyse der Artenzusammensetzung des Hordeetum murini in sieben europäischen Großstädten entlang eines West-Ost-Transektes: Ein Beitrag zur Charakterisierung der Stadtflora. *Phytocoenologia,* 23, 319-342.

Wittig R. & Sukopp H. (1998) Was ist Stadtökologie? In: *Stadtökologie. Ein Fachbuch für Studium und Praxis* (eds. Wittig R. & Sukopp H.). Gustav Fischer, Stuttgart, pp. 1-12.

Wright I.J., Reich P.B., Cornelissen J.H.C., Falster D.S., Groom P.K., Hikosaka K., Lee W., Lusk C.H., Niinemets U., Oleksyn J., Osada N., Poorter H., Warton D.I. & Westoby M. (2005) Modulation of leaf economic traits and trait relationships by climate. *Global Ecology and Biogeography,* 14, 411-421.

Zierdt M. (2006) Das graue Halle wird wieder bunt oder Warum die Flechten nicht mehr sterben. In: *Halle und sein Umland. Geographischer Exkursionsführer* (eds. Friedrich K. & Frühauf M.). Mitteldeutscher Verlag, Halle (Saale).

Zobel M. (1997) The relative role of species pools in determining plant species richness: an alternative explanation of species coexistence? *Trends in Ecology & Evolution,* 12, 266-269.

Appendix

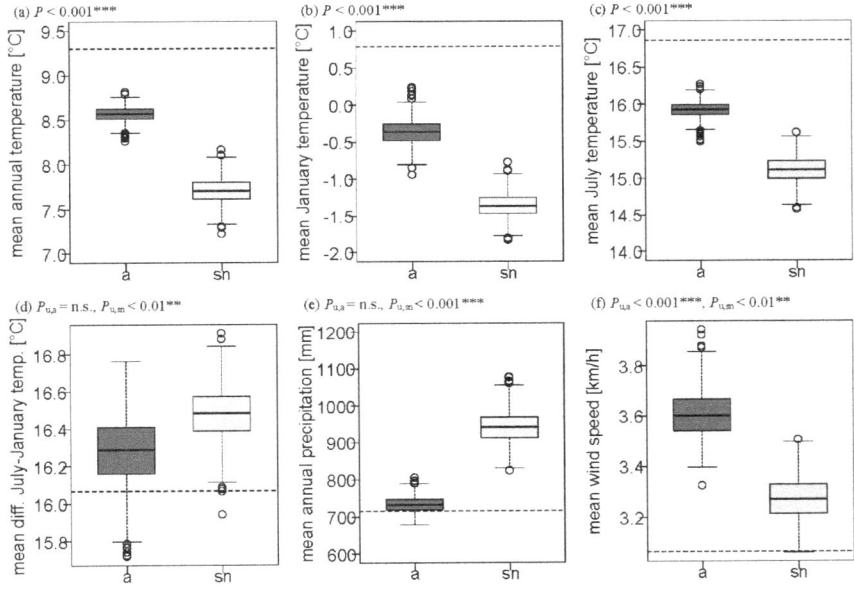

Figure A. 1 – Differences in climatic, topographic, edaphic, and geologic parameters between urbanized, agricultural and semi-natural grid-cells in Germany
(a) Mean annual temperature; (b) mean January temperature; (c) mean July temperature; (d) mean difference between July and January temperature; (e) mean annual precipitation; (f) mean wind speed; (g) number of geological patches; (h) number of geological types; (i) number of soil patches; (j) number of soil types;(k) mean height above sea level in urbanized (---), agricultural (*a*) and semi-natural (*sn*) grid-cells. See below for further information.

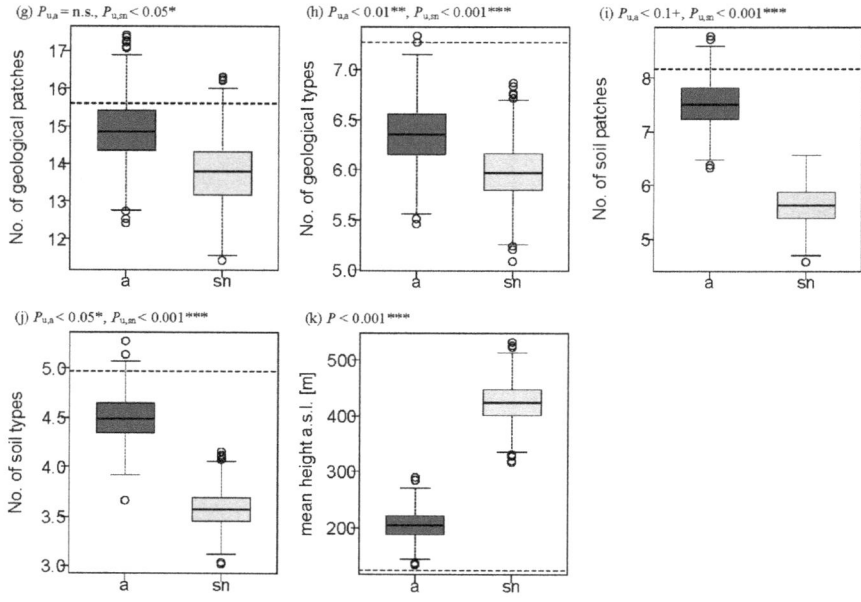

Figure A. 1. – continued

Futher explanation for figure A. 1.: Mean values for agricultural and semi-natural grid-cells are based on resampling (999*59 grid-cells; in accordance with the resampling method used in Chapters I and V; see *Materials and Methods* sections in these Chapters) and shown as dark grey (agricultural) and light grey (semi-natural) boxplots representing median (line), 25-75 % quartiles (boxes), ranges (whiskers) and extreme values (circles). Values for urbanized grid-cells are represented by a dashed line (---). If only one P-value is shown, it is valid for both the difference between urbanized and agricultural grid-cells and the difference between urbanized and semi-natural grid-cells. Otherwise, two P-values are shown ($P_{u,a}$ for difference between urbanized and agricultural grid-cells; $P_{u,sn}$ for difference between urbanized and semi-natural grid-cells).

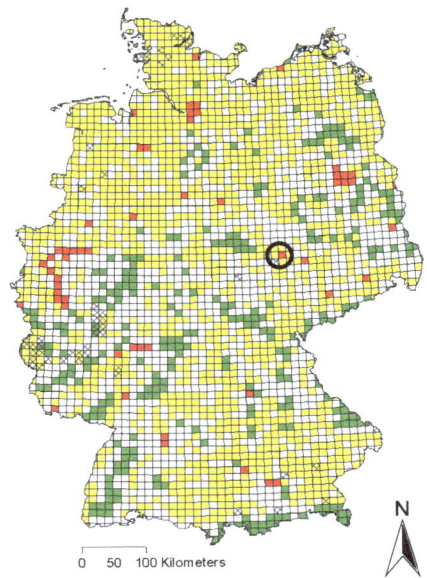

Figure A. 2 – Germany divided into grid-cells

Each grid-cell is sized *c*. 12 km × 11 km, i.e. *c*. 130 km² each. These cells are the study area of Chapters I, III and V. Red: urbanized grid-cells; yellow: agricultural grid-cells; green: semi-natural grid-cells; cross-hatched: grid-cells not used due to insufficient number of control species; white: grid-cells not meeting the selection criteria and thus not used (all as defined in Chapter I). The black circle marks the city of Halle, i.e. the study area of Chapters II and IV.

Table A. 1 – List of traits and trait states used for the analyses
Traits originate from BiolFlor (Klotz *et al.* 2002; Kühn *et al.* 2004b; http://www.ufz.de/
biolflor), LEDA (Kleyer *et al.* 2008; http://www.leda-traitbase.org/LEDAportal), SID
(Flynn *et al.* 2004), and Ellenberg *et al.* (2001). Abbreviations and descriptions are given.
'Chapter' shows in which analyses a trait (state) was used. Asterisks mark the trait state
used in the denominator of log-ratios in chapters I, II, and/or IV.

Trait	Trait state	Abbreviation	Description	Source	Chapter
Canopy height			Distance between the highest photosynthetic tissue and the base of the plant [m]	LEDA	III
Dispersal type	Anemo-chory*		Dispersal by wind*	LEDA	I-V
	Chamae-chory		Dispersal unit rolling over the soil surface caused by wind	LEDA	III
	Dysochory		Dispersal by scatter-hoarding animals	LEDA	I-III
	Endozoo-chory		Dispersal after digestion	LEDA	I-III
	Epizoochory		Adhesive dispersal	LEDA	I-III
	Hemerochory		Dispersal by man	LEDA	I-V
	Hydrochory		Dispersal by water	LEDA	I, II, III, V
	Zoochory (including dyso-, endo-zoo- and epizoochory)		Dispersal by animals	LEDA	IV, V
Ellenberg indicator values	Moisture value	Ellenberg_F	Showing the realized niche of a species with respect to moisture with values 1-12. Plants with low values can grow in extremely dry habitats; plants with high values can grow in ex-tremely wet habitats.	Ellenberg *et al.* 2001	III, IV

Trait	Trait state	Abbreviation	Description	Source	Chapter
Ellenberg indicator values *(continued)*	Nitrogen value	Ellenberg_N	Showing the realized niche of a species with respect to nitrogen with values 1-9. Plants with low values can grow in nitrogen-poor habitats; plants with high values can grow in nitrogen-rich habitats.	Ellenberg *et al.* 2001	III, IV
	Temperature value	Ellenberg_T	Showing the realized niche of a species with respect to temperature with values 1-9. Plants with low values can grow in cold habitats; plants with high values can grow in hot habitats.	Ellenberg *et al.* 2001	III, IV
Floristic status	Archaeophyte		Taxon introduced before the discovery of the Americas	BiolFlor	III-V
	Native		Taxon native (i.e. indigenous) to Germany	BiolFlor	III-V
	Neophyte*		Taxon introduced after the discovery of the Americas	BiolFlor	III-V
Floristic zone	Allrounder	FZ_allrounder	Origin in all climatic zones	BiolFlor	III
	Exratropical allrounder	FZ_extratropic	Origin in every but the tropic zone	BiolFlor	III
	(Sub-) meridional	FZ_meridional	Origin in the zone of evergreen broad-leaved and coniferous forests, summer-green dry forests, steppes and deserts	BiolFlor	III
	Temperate	FZ_temperate	Origin in the temperate zone with summer green deciduous forests	BiolFlor	III

Trait	Trait state	Abbreviation	Description	Source	Chapter
Floristic zone *(continued)*	Temperate and boreal/arctic	FZ_temperate-north	Origin in the temperate zone and the northern/southern taiga coniferous forests and/or tundras from the treelines polewards	BiolFlor	III
	Temperate-meridional	FZ_temperate-merid.	Origin in the temperate zone and the (sub-) meridional zone	BiolFlor	III
Flowering phenology *(continued)*	Pre-spring		Flowering in pre-spring	BiolFlor	IV
	Early spring		Flowering in early spring	BiolFlor	IV
	Mid spring		Flowering in mid spring	BiolFlor	IV
	Early summer		Flowering in early summer	BiolFlor	IV
	Midsummer		Flowering in mid-summer	BiolFlor	IV
	Early autumn		Flowering in early autumn	BiolFlor	IV
Leaf anatomy	Helomorphic		With aeration tissue in the root as adaptation to oxygen deficiency in swampy soils	BiolFlor	I-V
	Hydromorphic		Adapted to gas exchange in the water	BiolFlor	I, II, IV, V
	Hygromorphic*		Delicate plants of shade and semi-shade*	BiolFlor	I-V
	Mesomorphic		Without any characteristics, between scleromorphic and hydromorphic	BiolFlor	I-V

Trait	Trait state	Abbreviation	Description	Source	Chapter
Leaf anatomy *(continued)*	Scleromorphic		Firm and stiff leaves with thickened epidermis and cuticula but with mechanisms to promote water transport under beneficial conditions	BiolFlor	I-V
	Succulent		With water storage tissue and thickened epidermis and cuticula	BiolFlor	I-V
Leaf distribution along the stem	Regular	LD_regular	Leaves distributed regularly along the stem	LEDA, BiolFlor	III
	Rosette		Leaves concentrated near soil or water surface	LEDA, BiolFlor	III
	Hemirosette		Leaves arranged either scattered or tightly packed at the shoot	LEDA, BiolFlor	III
	Scarce	LD_scarce	Shoot scarcely foliated	LEDA, BiolFlor	III
Leaf dry matter content		LDMC	The ratio of dry leaf mass to fresh leaf mass; a measure of tissue density [mg/g]	LEDA	I, II
Leaf persistence	Evergreen		Leaves at all seasons often living more than one year	BiolFlor	I-V
	Spring green		Green from early spring to early summer; then usually decaying	BiolFlor	I-V
	Summer green		Green leaves only in the warm season	BiolFlor	I-V

Trait	Trait state	Abbreviation	Description	Source	Chapter
Leaf persistence *(continued)*	Overwintering green*		Leaves developing in autumn, overwintering green and decaying in spring and summer	BiolFlor	I-V
Life form	Hydrophyte		Resting buds are situated under water on the bed or in the mud	BiolFlor	I, II, IV, V
	Chamaephyte		Resting buds are situated on herbaceous or only slightly lignified shoots some centimeters above the soil surface protected by parts of the plant itself and/or by a snow cover	BiolFlor	I-V
	Geophyte		Resting buds are subterranean, often on storing organs protected within the soil	BiolFlor	I-V
	Hemicryptophyte		Resting buds are situated on herbaceous shoots close to the soil surface, protected by foliage or dead leaves	BiolFlor	I-V
	Phanerophyte		Resting buds are situated on (woody) shoots above the soil surface	BiolFlor	I-V
	Therophyte*		Summer annuals, which can only reproduce by means of generative diaspores	BiolFlor	I-V
Life span	Annual		The individual cycle lasts for a maximum of one year	BiolFlor	I, II, IV, V

Trait	Trait state	Abbreviation	Description	Source	Chapter
	Biennial		The plant grows for approximately one year vegetatively efore reaching the generative phase after which it completes its life cycle	BiolFlor	I, II, IV, V
	Pluriennial (summarizing mono- and polycarpic pluriennials)*		The plant grows longer than one year	BiolFlor	I, II, IV, V
Life span and clonality	Short lived (annual and biennial)		The individual cycle lasts for a maximum of two years	BiolFlor	III
	Pluriennial and non-clonal		Plants with an individual cycle of more than two years, reproducing by seeds	BiolFlor	III
	Pluriennial and clonal		Plants with an individual cycle of more than two years, reproducing vegetatively	BiolFlor	III
Long distance dispersal		LDD	Ability to transport dispersal unit over long distances	LEDA	III
Pollination type	Insect-pollination		Pollination by insects	BiolFlor	I-V
	Self-pollination (including cleistogamy, geitonogamy, pseudocleistogamy, selfing)		Spontaneous pollination within a flower or by a flower from the same plant	BiolFlor	I-V
	Wind-pollination*		Pollination by wind	BiolFlor	I-V
Seed mass			The air dried weight of a dispersule [mg]	LEDA, SID	III
Specific leaf area		SLA	The ratio of fresh leaf area to leaf dry mass [mm²/mg]	LEDA	I-IV
Type of reproduction	Only or mostly by seed		Plant reproduces by seed or spore, rarely vegetatively	BiolFlor	I-III

Trait	Trait state	Abbreviation	Description	Source	Chapter
	By seed and vegetatively		Plant is able to reproduce by seed and vegetatively	BiolFlor	I, II
	Only or mostly vegetatively*		Plant reproduces vegetatively, rarely by seed or spore	BiolFlor	I-III
UV-reflection of flowers	Yes*		UV reflectance pattern present	BiolFlor	I, II
	No		No UV reflectance patterns	BiolFlor	I, II

Table A. 2 – Number of species used per trait or trait state in the chapters of this book
See below for further explanation.

Number of species per Chapter

Trait	Trait state	I	II Protected areas	II 0.06 km²-plots	III	IV 1687-1689	IV 1721-1783	IV 1806-1856	IV 1857-1901	IV 1902-1949	IV 1950-1999	IV 2000-2008	V
Canopy height					1604								
Dispersal type	Anemochory	468	217	135	468	190	205	210	205	206	220	214	464
	Chamaecchory				34								
	Dysochory	168	117	81	168								165
	Endozoochory	406	253	168	410								403
	Epizoochory	780	404	253	781								772
	Hemerochory	510	295	241	510	223	244	263	272	264	309	304	504
	Hydrochory	253	136	81	256	118	123	135	137	128	135	130	251
	Zoochory					401	431	459	466	462	519	504	
Ellenberg indicator values	Moisture value				1516	517	561	602	596	567	633	605	
	Nitrogen value				1476	513	556	595	591	567	626	598	
	Temperature value				1341	434	473	504	510	487	544	524	
Floristic status	Archaeophyte				180	106	115	122	126	111	111	102	218
	Native				1273	599	646	681	660	633	655	627	2646
	Neophyte				259	6	16	26	39	49	131	128	655

Further explanation for table A.2.: Numbers for Chapter II are divided into numbers for protected areas and numbers for 0.06 km²-plots. Numbers for Chapter IV are divided into numbers per time period. Numbers for Chapter I and V differ slightly because for a small number of species there is no phylogenetic code available. Numbers for Chapters I and III differ because for Chapter III some traits could be complemented from other databases (Seed Information Database, and some traits from BiolFlor complemented from LEDA and vice versa), because only species with an available phylogenetic code were used, and because more rigorous restriction were made, namely omitting aquatic species and apomictic *Rubus* species.

Table continues on the following pages.

Table A.2. – continued

| Trait | Trait state | Number of species per Chapter | | | | | | | | | | | |
		I	II Protected areas	0.06 km²-plots	III	IV 1687-1689	1721-1783	1806-1856	1857-1901	1902-1949	1950-1999	2000-2008	V
Floristic zone	Allrounder				70								
	Exratropical. allrounder				446								
	(Sub-)meridional				107								
	Temperate				31								
	Temperate and boreal/arctic				36								
	Temperate-meridional				999								
Flowering phenology	Pre-spring					16	16	18	20	20	25	24	
	Early spring					74	79	84	85	83	90	89	
	Mid spring					119	134	133	128	133	140	133	
	Early summer					263	277	294	298	290	300	287	
	Midsummer					171	183	201	199	186	192	187	
	Early autumn					6	8	8	8	8	10	10	
Leaf anatomy	Helomorphic	448	156	70	208	162	172	181	178	152	148	137	447
	Hydromorphic	132	24	2	24	26	32	35	37	33	31	27	132
	Hygromorphic	325	135	74	226	101	112	118	127	137	148	138	325
	Mesomorphic	2210	657	449	1220	494	540	587	582	564	669	647	2202
	Scleromorphic	784	284	168	516	225	250	255	252	244	276	267	781
	Succulent	72	19	13	39	14	15	13	11	13	16	14	72

Table A2. – continued

Trait	Trait state	Number of species per Chapter												
		I	II Protected areas	0.06 km²-plots	III	IV 1687-1689	1721-1783	1806-1856	1857-1901	1902-1949	1950-1999	2000-2008	V
Leaf distribution along the stem	Regular				940								
	Rosette				110								
	Hemirosette				666								
	Scarce				26								
Leaf dry matter content		981	379	215									
Leaf persistence	Evergreen	887	232	137	456	205	223	236	235	234	252	240	883
	Spring green	63	27	2	43	20	25	23	28	28	32	32	63
	Summer green	1668	530	338	955	400	434	461	448	431	506	482	1655
	Overwintering green	256	106	77	178	94	102	113	120	106	107	104	256
Life form	Hydrophyte	148	32	5	27	33	40	43	45	42	39	35	148
	Chamaephyte	178	35	24	93	28	31	31	32	33	38	36	178
	Geophyte	438	111	40	247	107	120	122	120	119	135	128	438

Table A2. – continued

Trait	Trait state	I	II Protected areas	0.06 km²-plots	III	IV 1687-1689	1721-1783	1806-1856	1857-1901	1902-1949	1950-1999	2000-2008	V
Life form *(continued)*	Hemikryptophyte	1955	581	344	1022	493	517	553	543	526	563	541	1948
	Phanerophyte	746	148	118	220	46	57	60	61	72	113	110	744
	Therophyte	722	228	167	421	190	209	234	249	214	233	220	722
Life span	Annual	748	240	175		199	218	244	258	224	241	228	748
	Biennial	266	76	67		67	74	84	82	79	87	79	266
	Pluriennial	2904	703	386		536	587	611	600	600	691	665	2895
Life span and clonality	Short lived				536								
	Pluriennial non-clonal				558								
	Pluriennial clonal				757								
Long distance dispersal					1488								
Pollination type	Insect-pollination	2676	690	424	1297	533	586	620	620	605	691	663	2676
	Self-pollination	2119	570	344	905	385	419	454	461	442	490	471	2119
	Wind-pollination	741	233	141	364	177	192	210	205	197	211	203	741

Table A2. – continued

Trait	Trait state	Number of species per Chapter											
		I	II Protected areas	II 0.06 km²-plots	III	IV 1687-1689	IV 1721-1783	IV 1806-1856	IV 1857-1901	IV 1902-1949	IV 1950-1999	IV 2000-2008	V
Seed mass		1168			1305								
Specific leaf area		1702	520	301	1287	433	465	496	494	475	504	486	
Type of reproduction	Only or mostly by seed	1798	525	364	1673								
	By seed and vegetatively		361	187									
	Only or mostly vegetatively	168	43	12	764								
UV-reflection of flowers	Yes	349	192	125									
	No	533	299	191									